T0306143

Mastering Project Leadership

This is a collection of essays from key researchers in the field of project management who describe what they feel are the most impactful findings from research. In the challenging and competitive world of project management, project managers need all the insight they can get. Leading researchers share what they believe are the most important findings from the research being done today. These cover such pressing topics confronting project managers, including hybrid methodologies, schedule overrun, schedule estimation, project efficiency, and managing local stakeholders. Highlights include the following:

- Jeff Pinto and Kate Davis explore the "Normalization of Deviance" (NoD) phenomenon within various organizational settings, focusing on projects. NoD involves the gradual acceptance of deviant practices, diverging from established norms, and often leading to detrimental outcomes.

- Francesco Di Maddaloni investigates how local communities' stakeholders are perceived, identified, and categorized by project managers in major public infrastructure and construction projects (MPIC). His chapter helps project managers to have a better understanding of a more inclusive and holistic approach to engage with a broader range of stakeholders.

- Lavagnon Ika, Peter Love, and Jeff Pinto suggest that error and bias combine to exact a toll on major projects and offer theoretical insights, and outline practical recommendations for project managers.

- Jonas Söderlund offers managerial guidelines for leveraging deadlines as powerful tools for generating project success.

- Pedro M. Serrador looks at empirical studies that link planning quality to project success, emphasizing its importance. He also discusses the downsides of excessive planning, particularly in dynamic environments and research and development projects.

Mastering Project Leadership
Insights from the Research

Edited by
Pedro M. Serrador

CRC Press
Taylor & Francis Group
Boca Raton London New York

CRC Press is an imprint of the
Taylor & Francis Group, an **informa** business
AN AUERBACH BOOK

First edition published 2025
by CRC Press
2385 NW Executive Center Drive, Suite 320, Boca Raton FL 33431

and by CRC Press
4 Park Square, Milton Park, Abingdon, Oxon, OX14 4RN

CRC Press is an imprint of Taylor & Francis Group, LLC

ISBN: 978-1-032-47332-1 (hbk)
ISBN: 978-1-032-47256-0 (pbk)
ISBN: 978-1-003-50265-4 (ebk)

DOI: 10.1201/9781003502654

Typeset in Minion
by MPS Limited, Dehradun

Contents

Editor

Pedro M. Serrador, PhD, P.Eng., PMP, ITIL, PMI-ACP is a writer and researcher on Agile, AI and management topics. He is an adjunct professor at Northeastern University, Boston and University of Toronto, Canada.

He is also the owner of Serrador Project Management, a consultancy in Toronto, Canada. He specializes in technically complex and high risk programs and projects, vendor management engagements, and tailoring and implementing project management methodologies. He has worked on projects in the financial, telecommunications, utility, medical imaging, and simulations sectors for some of the world's largest companies for more than 25 years. He also takes time to invest in early stage startups.

He an author of books and articles on project management and is also a regular speaker at conferences. He was the recipient of the PMI 2012 James R. Snyder International Student Paper of the Year Award, the Major de Promotion Award for best PhD Thesis 2012–2013 from SKEMA business school and a 2022 Project Management Journal most cited paper of the year award.

Contributors

Robert E. Bierwolf
Foundation Center of Technology
   and Innovation Management
   (CeTIM)
Gateway Office
National Organization for
   Development
Digitization and Innovation (ODI)
Ministry of the Interior and
   Kingdom Relations
Netherlands

Chantal Cantarelli
Cranfield School of Management
Cranfield, United Kingdom

Kate Davis
Cranfield School of Management
Cranfield, United Kingdom

Nathalie Drouin
ESG UQAM
Montreal, Quebec, Canada

Gabriela Fernandes
University of Coimbra, Faculty of
   Sciences and Technology
Coimbra, Portugal

Pieter H.A.M. Frijns
Amsterdam Business School,
   Faculty of Economics
Vrije Universiteit Amsterdam
Amsterdam, Netherlands

Andrew Gemino
Beedie School of Business
Simon Fraser University
Vancouver, British Columbia,
   Canada

Lavagnon Ika
Telfer School of Management
University of Ottawa
Ontario, Canada

Peter Love
School of Civil and Mechanical
   Engineering
Curtin University
Perth, Australia

Francesco Di Maddaloni
The Bartlett School of Sustainable
   Construction
University College
London, United Kingdom

Carl Marnewick
University of Johannesburg
Johannesburg, South Africa

Ralf Müller
BI Norwegian Business School
Oslo, Norway

David O'Sullivan
University of Galway
Galway, Ireland

Jeffrey K. Pinto
Black School of Business
Penn State
Erie, Pennsylvania, United States

Blaize Horner Reich
Beedie School of Business
Simon Fraser University
Vancouver, British Columbia,
   Canada

Shankar Sankaran
School of the Built Environment
University of Technology Sydney
Sydney, Australia

Pedro M. Serrador
Northeastern University
Toronto, Ontario, Canada

Jonas Söderlund
Linköping University
Linköping, Sweden

Neil Turner
Cranfield School of Management
Cranfield, United Kingdom

Rodney Turner
School of Civil Engineering
University of Leeds
Leeds, United Kingdom

Introduction

The idea for this book came at a project management conference a couple of years ago. After the conference, some colleagues and I were discussing over dinner the fact that much of the research that goes on in academia never reaches practitioners: those project managers working day in and day out to deliver successful projects. We discussed how it was a shame there was no way to get good project management research out into the business word.

On my flight back home, I thought about some books I had read that brought research to the public and thought that was something I could help do for project management. That was the genesis of the idea for this project.

I sent some feelers out to some of the most consequential academics I knew in the project management field. I asked them if they would be interested in writing about some of their most interesting research for a wider project manager audience. I was happy to see that there were generally positive responses. And I am now happy to have some of the leading names in the world of project management research contributing to this book.

Some of the article still retain some of the statistics and descriptions of methods that were used in the original academics articles, but not all. I hope this gives readers some insight and understanding into how academic research is conducted. And I hope it is not distracting or confusing. But readers are free to skip over those sections; though I suggest digging in and spending some time understanding the statistics.

I hope project managers will find these articles interesting and perhaps find ideas that help them in their often challenging jobs. At the very least, I hope these articles get you thinking about the nature of and challenges inherent in project management.

Pedro M. Serrador
Toronto, Ontario, Canada

1

When Our Project Culture Fails Us[1]: The Normalization of Deviance Trap

Kate Davis[1] and Jeffrey K. Pinto[2]
[1]Cranfield School of Management, Cranfield, United Kingdom
[2]Black School of Business, Penn State, Erie, Pennsylvania, United States

On 13 January 2012, the Carnival cruise ship Costa Concordia crashed into rocks off Giglio Island, on the Italian coast, leading to the deaths of 32 of the 4,252 passengers on board. The sinking has been attributed to gross negligence of the captain and crew. Captain Francesco Schettino first decided to go on an unapproved course because it was a tradition for cruise ships to pass the island closely, even if that meant steering out of approved shipping lanes into dangerous, shoaling water. Prosecutors found that the ship was cruising too close to the island in a "ship salute" publicity stunt before it rammed into the submerged rocks. According to later reports, Carnival's directors "not only tolerated, but promoted and publicized the risky ship salutes off Giglio and other tourist sites as a convenient, effective marketing tool" (Vogt, 2013). In other words, passing closely created a spectacle for the people on the shore. This behavior was the norm, with each captain deviating from the approved path.

This example illustrates a troubling organizational phenomenon, in which members of the firm develop a tolerance for deviation, or the willingness to violate or ignore accepted standards of behavior. The term itself, "normalization of deviance" was first coined in the aftermath of the space shuttle Challenger disaster, which revealed a series of missteps, flawed assumptions, and a culture of risk-taking. "Social normalization of deviance means that people within the organization become so much accustomed to a deviant behavior that they don't consider it as deviant, despite the fact that they far exceed their own rules for the elementary safety" (from Villeret interview with Vaughan, Villeret, 2008). More insidiously, previous work has found that people grow more accustomed to the deviant behavior the more it occurs, thus desensitizing members of the organization to the norms they are violating or even the fact that their violations matter. In practical terms, it represents behaviors that can lead to a number of significant problems with project development and delivery, as cultural failures and flawed governance permit and normalize the patterns of destructive actions by key stakeholders that are counter-productive to organizational expectations.

Put simply, normalization of deviance suggests that over time, the unexpected becomes the expected, which becomes the accepted (Pinto, 2006). Perhaps a construction safety step is ignored "just this one time" in order to speed up project completion and when no ill effects occur, workers begin to believe that the safety step wasn't really even necessary in the first place. So, the next time a similar situation occurs, workers feel increasingly comfortable with bypassing the safety step, until the development process simply drops such "time-consuming" delays to the project. In this way, the "one time" decision to ignore a safety check (the unexpected) gradually becomes expected and eventually, accepted behavior. Thus, one phenomenon of this normalization of deviance is that while a series of behaviors may appear deviant to people outside the organization, for personnel within the firm, the deviance often goes unrecognized; that is, it is simply assumed to be normal occurrence. It is usually only with hindsight that people within an organization can realize that their seemingly "normal" behavior was, in fact, deviant (Vaughan, 1999, 2004). Or, to put it bluntly, as one project manager related to us about a serious accident on a job site, in which the firm knowingly relaxed safety standards to speed the work, "It seemed like a good idea, until it wasn't."

This chapter will examine normalization of deviance (NoD) within project settings, to better understand why the practice frequently occurs, the ways in which it most commonly manifests itself, and how project team members and managers can begin to take remedial action to identify and address this issue, before disasters take place. In practical terms, NoD represents behaviors that can lead to a number of significant problems with project development and delivery, as cultural failures and flawed governance permit and normalize the patterns of destructive actions by key stakeholders that are counter-productive to organizational expectations.

NORMALIZATION OF DEVIANCE – WHY DOES IT OCCUR?

The first point to recognize is that NoD behaviors are quite common; examples abound in diverse industries and occupations. Everything from nursing and healthcare provision to construction and workplace safety to auditing and finance, media, the list of organizations and project settings in which examples of NoD behaviors are found is wide and cuts across corporate hierarchies, disciplines, and training/background. In simple terms, NoD happens as a result of failures. These failures happen within the governance and control systems of organizations, reinforced by counter-productive or perverse reward systems, cultures that stifle dissenting views, as well as the challenges that operating in project teams naturally brings. We know that the unique nature of project work is characterized by a variety of difficulties, in developing products or services while addressing a litany of often-competing stakeholder, safety, budgetary, schedule, and quality expectations, all while working with temporarily assembled teams and emergent leadership. Because of the wide variety of external and internal pressures on perform-ance, these teams may not have the familiarity and comfort level to develop a supportive, positive culture, and instead feel the pressures from outside groups and challenges to get the work done, "any way necessary."

An important cause of deviant behavior is the organizational culture within which workers perform their duties. Organizational culture has been defined in various ways but one of the best is offered by Ed Schein (1985: p. 30), who defined culture as: "the pattern of basic assumptions that the group has invented, discovered, or developed in learning to cope with its problems of external adaptation and internal integration, and has worked well enough to be considered valid and therefore, taught to new members." We can see from this definition that cultures form around basic ideas of expected behavior (how we should act), ways we communicate (patterns of language and meaning), and how we should relate to each other and the external environment (how we interpret events and act accordingly). An understanding of culture is a critical clue to the pressures felt by organizational members to engage in or avoid deviant behaviors; for example, if the common understanding is that "everyone does it," members of the workgroup or larger organization often find it easier to rationalize "going along to get along" rather than risk social or professional ostracism. As a result, a first step to understanding both the frequency of deviant behavior and the degree to which it is socially normalized lies in understanding the cultural forces that can subtly, but powerfully, affect each member of the organization.

Another part of the challenge in recognizing and addressing NoD is the role that the "gradualism" phenomenon plays in promoting these concerns. As Starbuck and Milliken (1988) have noted, getting used to "deviance" behavior occurs as a process of steps, often over an extended period. Unacceptable behaviors may not occur all at once, but rather, may serve as the summation of multiple decisions made or avoided, with no visible or discernible negative effects. Thus, the potential for catastrophe is never envisioned as an option until it occurs. The consequence of success (or an absence of failure), according to these authors (Starbuck and Milliken, 1988: p. 322), results in a general attitude of "complacency versus striving, confidence versus caution, inattention versus vigilance, routinization versus exploration, habituation versus novelty. Successes foster complacency, confidence, inattention, routinization, and habituation; and so human errors grow increasingly likely as successes accumulate." In his assessment of the causes of the Challenger Space Shuttle disaster, for example, noted physicist Richard Feynman [quoted in Vaughan, 1996: p. I-148] observed: "... [After] each successful flight, NASA's managers thought 'We can lower our standards a bit because we got away with it last time'." Another example occurred with Boeing's 2018 introduction of their widely used 737, upgraded as the 737 Max. In the aftermath of two fatal accidents and the deaths of over 300 people, the company has been charged with negligence through pushing these design and software upgrades too rapidly, even though they fundamentally changed the flight characteristics of the aircraft. In this case, commercial pressures were assumed to trump technical concerns, as the changes were rationalized as simple "upgrades" to a proven airframe. Normalization of deviance represents a cultural attitude that consciously creates conditions in which mistakes are made; in effect, it provides a perfect petri dish environment for corporate (or project) misbehavior.

WHAT ARE SOME WAYS WE SEE NORMALIZATION OF DEVIANCE IN PROJECTS?

We first need to understand that NoD behaviors are taught to newcomers in our organizations through a series of three practices:

institutionalization, socialization, and rationalization. Each of these steps is critical in first introducing and then ultimately, cementing these deviant practices in place for all project team members.

Institutionalization exposes newcomers to deviant behaviors, often performed by authority figures, and explains those behaviors as organizationally normative. When new members of project teams are first assigned, they are quickly immersed in the rules (written and unwritten) that govern project activities (Argyris and Schon, 1978). Because new members may be aware of the "right way" to perform tasks, institutionalization processes are intended to show them "how we do it here" in order to quickly forestall their objections should these behaviors seem unethical or unsafe.

Socialization, which is usually reinforced by a variety of rewards and punishments, aims at determining whether the newcomer will or will not join the group by adopting the group's deviant behaviors. This step is where governance and operating culture most directly collide, as new members are exposed, through experiencing critical incidents and subsequent rewards or sanctions, to expected behaviors and are at this point presented with the implicit choice of joining in to get along (Gaim et al., 2021), or risking isolation and social ostracism by not submitting to the cultural norms of the project team.

Rationalization enables organizational members to convince themselves that their deviances are not only legitimate, but acceptable and perhaps even necessary. The gradualism at work in NoD is most often demonstrated as part of the rationalization step. Repeated missteps or deviations from accepted operating norms and principles are ignored to the point where they become institutionalized and accepted – even expected – on the part of project team members.

Previous research has examined the specific examples and most common ways in which NoD behaviors occur in projects (Davis and Pinto, 2022). Let us consider each of these settings and the misbehaviors that can happen:

1. *Project proposals and strategic misrepresentation* – The term "strategic misrepresentation" refers to the deliberate use of misleading or false information by key project stakeholders (customers or contractors). One avenue for NoD lies in the tactics employed by firms to win competitive bids or as part of scope negotiations, often through falsifying pertinent information, minimizing risks, making

unrealistic project delivery promises, and so forth. Knowing full well that in many cases, these initial promises, though perceived as crucial for winning the business, are based on well-understood falsehoods, project organizations tacitly (and sometimes overtly) encourage these behaviors. As a recent example, an investigation of wide-spread corrupt practices in the Canadian construction industry in 2011 identified bid-rigging and price fixing in the awarding and management of public contracts. The commission announced that the corruption and collusion were "far more widespread than originally believed" in the construction sector (Saint-Martin, 2015) and was considered the "usual" way of managing public contracts (Courtois and Gendron, 2017).

2. *Client/contractor relationships* – A fascinating feature of many client/contractor relationships is that they often follow a common "rival camps" dynamic. In our own experiences, many of us have seen that rather than explore opportunities to create partnerships and open communications, critical project information is often hoarded and either misused or doled out selectively. Clients may not trust contractors, convinced they are being lied to in order for the contractor to maximize their profits at the customers' expense (Pinto, et al, 2009). Left to their own devices, as a result, the emergent pattern among clients and contractors is often one of indirect conflict and opportunism.

3. *Planning and scheduling dynamics* – By scheduling dynamics, we are referring to the wide variety of problems that often occur during the project planning and scheduling cycle. To create accurate schedules, it is necessary for project managers to have full information and a constructive, trusting relationships with senior managers. When a project manager is asked to develop a schedule, there is an implicit assumption that estimates will be in good faith and the resulting project plan reflects a reasonable path to completion. However, it may not be this way at all. Problems in perception, false manipulation, or hijacking the planning process outright, and pressures that senior executives often bring to bear to artificially adjust the schedules routinely pit top management against the project manager (Pinto, 2022).

4. *Workplace safety* – Another critical example of NoD behaviors occurs when organizations gradually allow safety standards to relax while pursuing project outcomes. That is, as the original example in

this chapter typifies, there are any number of projects in the construction industry, for example, that fail to enforce safety standards (Anderson et al., 2018; Smith, 2019). Although everyone – site workers and management – is aware of unsafe practices and fully recognizes that such behaviors should be avoided, there is often an unspoken sub-text accompanying these prohibiting rules in which it is not only possible but often expected that safety rules can be relaxed or ignored.

ADDRESSING NORMALIZATION OF DEVIANCE – WHERE TO START

Hopefully, we have made clear from the above discussion that NoD behavior occurs for a variety of reasons, some understandable if not excusable. Further, our past research identified some of the most common ways in which project teams and their organizations are affected by NoD. So, based on this background, the follow-on question has to be asked: what can be done about it? In this section of the chapter, we would like to suggest some methods by which firms can begin to recognize and address the conditions "on the ground" that promote NoD and how to remediate these practices, hopefully before tragic consequences occur. The nature of improving the cultural setting of the organization involves: 1) recognizing the degree to which normalization of deviance operates within our organizations and 2) developing methods for exposing, addressing and remediating these behaviors.

1. **Establish standards for acceptable behavior** – The first step in the process of addressing NoD is to recognize that ambiguity is not your friend; the more latitude and flexibility (or murkiness) in how behavior is controlled and rewarded, the greater the likelihood that workers will interpret a "lack of controls" as a willingness to "ignore controls."

2. **Take a hard, honest look at current practices** – Remember that the point about NoD behaviors is that they are widely engaged-in; *these are not hidden practices*, but widely recognized and accepted (or, at least, winked at). As a result, with a little prodding on the part of key organizational members, it is usually easy to come up

with standard behaviors, both positive and potentially destructive. It is also useful to consider using leaders outside of the work group to break loose the logjams of established thought processes. Groupthink can have the effect of allowing groups to jointly reassure each other that behaviors are okay but bringing in some outside perspective is often necessary to change the attitudes and perspective of those caught in these self-defeating cycles.

3. **Link these behaviors to outcomes** – Project team members have to see the link between their deviant practices and real negative outcomes for themselves, their project, and their organization. As long as their behaviors are not seen as "critical" or effectively linked to a negative outcome, it is hard to make the case that new standards are needed to replace the old ones. On the other hand, if management can show clear cause and effect ("You allowed your workers to ignore safety harness requirements and that led to a scaffolding accident"), it is easier for project team members to buy in to the need to alter current attitudes and behaviors.

4. **Reinforce through modified reward systems** – We need to understand that NoD practices become embedded in the culture because they are, in some manner, rewarded. When we promote or offer bonuses for rapid project completion, are we sending the unintended signal that all we care about are results, rather than results *and* process? One of the oldest laws of human behavior is that we tend to get the behaviors that we are reinforcing, either intentionally or unintentionally. A critical step is to restructure reward systems so that employees can see past simple task completion (and the unspoken corollary, "at any cost") to a clearer idea of not just *what* to do *but* how to get it accomplished.

5. **Be public and loudly present** – We suggested at the beginning of the chapter that NoD occurs as the result of a failure of organizational culture and governance systems. The challenge with changing a culture is that these are, by definition, the unwritten rules of the game. Simply adding a new poster to the wall of an employee break room will not get the changes you are looking for. On the other hand, the public and clear linking of rewards and/or sanctions for specific behaviors, along with a message that new attitudes are required and compliance is mandatory, makes it clear in peoples' minds that this is not some temporary quick-fix or PR stunt. We all prefer to reward good

behavior, but we have to be equally willing to punish violations. In either case, the more public we are with these responses, the better.

CONCLUDING THOUGHTS

This chapter has addressed a difficult subject, the way in which organizational reward systems, culture, and (if we are honest) our own willingness to look the other way at times have contributed to the normalization of deviance and the potentially disastrous consequences of these actions. We are not suggesting that all firms suffer from these actions on a broad scale, but it is fair to state that the potential exists for such misbehavior in all firms. Project organizations can provide a unique setting that makes us more willing to look the other way in an effort to (as "Larry the Cable Guy" says) "git 'er done!" Successful projects are not just about outcomes; they depend on ethical and rational practices, as well. Combining a desire to use best practices to create best outcomes is our means villeretfor avoiding the damage that normalization of deviance causes.

NOTE

1 Portions of this chapter was adapted from: Davis, K. and Pinto, J.K. (2022), The corruption of project governance through the normalization of deviance, *IEEE Transactions on Engineering Management*, (in press); and, Pinto, J.K. (2014), Project management, governance, and the normalization of deviance, *International Journal of Project Management*, 32, 376–387.

REFERENCES

Andersen, L.P., Nørdam, L., Joensson, T., Kines, P. and Nielsen, K.J. (2018). Social identity, safety climate and self-reported accidents among construction workers, *Construction Management and Economics*, 36, 22–31.

Argyris, C. and Schon, D.A. (1978). *Organizational Learning: A Theory of Action Perspective*. Reading, MA: Addison-Wesley.

Courtois, C. and Gendron, Y. (2017). The "normalization" of deviance: A case study on the process underlying the adoption of deviant behavior, *Auditing: Journal of Practice & Theory*, 36(3), 15–43.

Davis, K. and Pinto, J.K. (2022). The corruption of project governance through the normalization of deviance, *IEEE Transactions on Engineering Management*, (in press), DOI 10.1109/TEM.2022.3184871.

Gaim, M., Clegg, S.R., and Cunha, M.P.E. (2021). Managing impressions rather than emissions: Volkswagen and the false mastery of paradox, *Organizational Studies*, 42, 949–970.

Pinto, J.K. (2022). No project should ever finish late (and why yours probably will, anyway), *Engineering Management Review*, 50, 181–192.

Pinto, J.K. (2014). Project management, governance, and the normalization of deviance, *International Journal of Project Management*, 32, 376–387.

Pinto, J.K., Slevin, D.P., and English, B. (2009). Trust in projects: An empirical assessment of owner/contractor relationships, *International Journal of Project Management*, 27, 638–648.

Pinto, J.K. (2006). Organizational governance and project success: Lessons from Boston's Big Dig. Presentation at: Concept Symposium — Principles of Governance of Major Investment Projects, Trondheim, Norway.

Saint-Martin, D. (2015). Systemic corruption in an advanced welfare state: Lessons from the quebec charbonneau inquiry, *Osgoode Hall Law Journal*, 53(1), 66–106.

Schein, E.H. (1985). Organisational Culture and Leadership. San Francisco: Jossey Bass.

Smith, S.D. (2019). Safety first? Production pressures and the implications on safety and health, *Construction Management and Economics*, 37, 38–242.

Starbuck, W.H. and Milliken, F.J. (1988). Challenger: Fine-tuning the odds until something breaks, *Journal of Management Studies*, 25, 319–340.

Vaughan, D.S. (1996). *The Challenger Launch Decision: Risky Technology, Culture, and Deviance at NASA*. Chicago, IL: University of Chicago Press.

Vaughan, D.S. (1999). The dark side of organizations: Mistakes, misconduct, and disaster, *Annual Review of Sociology*, 25, 271–305.

Vaughan, D.S. (2004). Organizational rituals of risk and error. In: Hunter, B., Power, M. (Eds.), *Organizational Encounters with Risk*. Cambridge University Press, New York, pp. 33–66.

Villeret, B., (2008). Interview: Diane Vaughan. *Consulting News Line*, May, http://www.consultingnewsline.com/Info/Vie%20du%20Conseil/Le%20Consultant%20du%20mois/Diane%20Vaughan%20(English).html.

Vogt, A. (2013). Costa Concordia: UK directors to be named responsible for capsize. *The Week*, Jan 9, http://www.theweek.co.uk/europe/costa-concordia/50908/costa-concordia-uk-directors-be-named-responsible-capsize

2

Balanced Leadership: Making Use of All Leadership Skills in the Project Team

Ralf Müller[1], Nathalie Drouin[2], and Shankar Sankaran[3]
[1]BI Norwegian Business School, Oslo, Norway
[2]ESG UQAM, Montreal, Quebec, Canada
[3]School of the Built Environment, University of Technology Sydney, Sydney, Australia

INTRODUCTION

Much is written on leadership – the interpersonal social process of guiding people. The most popular theme within the scope of this definition of leadership is probably *leadership styles*, which describe the particular ways leaders interact with their followers (Thite, 2000). These styles may describe behavior directly, like leaders asking followers for input when using democratic styles, giving orders when using autocratic styles, or setting motivators when using transactional styles. Alternatively, leadership styles may indirectly describe a behavior, such as through an underlying value system. For example, in the case of servant leadership, the leader prioritizes the team and the organization over his or her personal objectives, resulting in a particular interaction between leader and follower. The choice of leadership style depends on the situation. Using a democratic style in an emergency can be just as inappropriate as an autocratic one when creativity and innovation is required. Hence, choosing the right leadership style in a given situation is important for successful leadership. Many publications address this. For example, Goleman, Boyatzis, and McKee (2002) take the perspective of the contemporary social and emotional intelligence school of leadership and suggest democratic styles when consensus is needed, coaching styles when people's performance needs improvement, or commanding styles in cases of crises.

An underlying assumption in the leadership styles literature is that there is a leader. In other words, a source or origin of leadership which exercises a leadership style. Traditionally, the literature distinguishes between leadership by a formally appointed leader, also known as *vertical leadership* (Turner & Müller, 2005), and leadership stemming from within the team, which is often referred to as *team-based leadership* (Pearce & Conger, 2003). Vertical leadership is hereby described as the leadership exercised by the person considered to be the leader, often the manager of the project. Team-based leadership is often distinguished into a) *shared leadership,* where the team decides whom to follow, and b) distributed leadership, where the team interacts, maybe also with outsiders, and through the exchange of opinions, discussions, and new perspectives, leadership emerges as an outcome of the team's interaction. We call these different origins of leadership the *leadership approach.* Within each of these leadership approaches, all leadership styles are possible. However, recent research has shown that vertical and team-based leadership approaches are insufficient to explain leadership in projects because of the dynamics in project realities.

In probably the largest study ever done in project management research and funded by the Project Management Institute (PMI), about 25 researchers, organized in nine country teams all over the world, conducted approximately 300 interviews in more than 80 in-depth case studies to develop a better understanding of the dynamics between vertical and team-based leadership and subsequently validated their findings using a global survey. The study's results, described among others in (Müller, Drouin, & Sankaran, 2021), show that the existing leadership approaches are found in projects. However, further approaches not described in the existing literature were also crucial for project success. These are horizontal and balanced leadership. The former applies when the project manager is not the expert in a particular skill area, but a skilled leader is required to move the project forward. The latter is applied to ensure that the best possible leader is in charge of leadership at any time in the project. These two leadership concepts complement the existing dichotomy of team-based and person-based leadership and explain how leadership authority is dynamically assigned to people in situational contingency in projects. We now discuss these two approaches in detail.

HORIZONTAL LEADERSHIP

Horizontal leadership happens when the project manager appoints a team member as a horizontal leader to guide the project through a particular situation. This appointment often happens when the project manager is not an expert in the subject area of the issue at hand, but the appointed horizontal leader is. For example, a database specialist in an IT project, whose design skills are crucial for turning around a project in crisis and with unhappy customers. Or an industry specialist of a seller organization who needs to consult the buying organization's business managers about adjusting the project outcome to emerging trends in the market. A characteristic of horizontal leadership is, besides the appointment of a team member, that the project manager subordinates to the appointed horizontal leader for the time of the appointment. At the same time, the project manager fulfills a governance role by observing the horizontal leader to ensure the best possible outcome for the project. This distinguishes horizontal leadership from the traditional task delegation, where the project manager would continue to be the manager while the delegated team member works on solving the problem.

During the appointment, the horizontal leader applies the leadership style he or she deems appropriate. Hence, all leadership styles are possible in a horizontal leadership approach.

As an intermediate between vertical and team-based approaches, horizontal leadership complements the existing leadership approaches of vertical, shared, and distributed leadership.

This appointment of horizontal leaders does not come "out of the blue." The above-mentioned study showed that preparations for horizontal leadership start early on in the project and are described as part of the balanced leadership framework below.

BALANCED LEADERSHIP

Earlier in this chapter, we said leadership styles are chosen depending on the situation in the project. The same applies to leadership approaches. For example, many projects require creativity and innovation in the early phases. Here is where team-based leadership allows for using the team's combined intellectual power. During the planning phase, central coordination is often required. The project manager uses vertical leadership, potentially allowing subgroups to develop their plans using team-based approaches but feeding their plans back into the vertically led planning phase. Similarly, issues encountered during the implementation stage may be solved through horizontal leadership or team-based approaches. Later, the project manager will most likely use vertical leadership to ensure lessons learned and other close-out tasks are executed. Hence, the choice of leadership approach continuously bounces back and forth between vertical, horizontal, and team-based approaches in situational contingency. Finding the right balance between situational demands and chosen leadership approach is described as *balanced leadership*. In other words, balanced leadership *"emerges from the dynamic, temporary, and alternating transitions between vertical, shared, distributed, and horizontal leadership for the accomplishment of desired states in, for example, task outcome, or the entire project"* (Müller et al., 2021, p. 10).

Balanced leadership and its core task of choosing the most appropriate leadership approach is made up of five events, which start early on in the project. The events need not necessarily occur as sequentially as described below. Sometimes they are nested in each other. However, for balanced leadership to happen, all five events must be completed. The events are:

- *Nomination.* This is when team members are appointed to the project. Here the project manager anticipates possible issues during the project and identifies the required skills to solve them. Once the required skills are known, the project manager aims to influence candidate selection to get as many of the anticipated skills into the project as possible.
- *Identification.* This is when the project manager singles out the existing team members' particular skills and judges their ability to act as temporary leaders.
- *Selection.* This is when the project manager empowers an individual as a horizontal leader or a sub-team to act as a team-based leader.
- *Horizontal or team leadership and its governance.* This is when the appointed individual or team accepts the appointment and takes on leadership accountability. The project manager temporarily

subordinates to the appointed leader but governs the appointed leader simultaneously during the time of the appointment.

- *Transition.* This is when the temporary leader's appointment ends after task completion or upon decision by the project manager.

The above shows that balanced leadership pervades the entire project, from the early nomination to the appointment of individuals or teams as temporary leaders until the end of their assignment. However, balanced leadership can only happen when the project manager recognizes its value and allows it. Some project managers are reluctant to give away their leadership authority. In these projects, the risk is high that sub-optimal leadership jeopardizes project results.

COORDINATION

One may ask how the above events are coordinated between the project manager and the team. Through the study, we learned that coordination takes place through a common understanding of the project manager and the team in respect of three facts of the project:

- *Who is currently empowered to lead?* This is needed to avoid clashes and conflicts between team members who may think people other than the nominated leaders should be appointed.
- What level of self-management does the appointed leader show? Self-management refers to the extent an appointed person is able to fullfil the assigned task without help from others. Suppose the appointment is mainly to give the person a chance to learn how to lead (i.e., low self-management capabilities). In that case, the team typically interacts supportively and consultatively with the appointed leader. Suppose the person is appointed because of high technical skills in solving the problem at hand (high self-management). In that case, the team typically follows the suggestions of the appointed leader without much feedback on leadership style and interaction.
- What skills are required at what time in the project, and which skills are available at these times? This shared mental model of skills requirements helps to understand who might be appointed as a

temporary leader at later stages in the project or even if there is a possibility that externals will lead the project temporarily.

These questions are answered, updated, and shared in the regular project team meetings. For balanced leadership to work, it is important that everyone in the team understands the answers to these questions to achieve transparency in leadership decisions and avoid conflicts (Drouin, Müller, Sankaran, & Vaagaasar, 2018; Drouin, Vaagaasar, Sankaran, & Müller, 2021).

CONCLUSION

This chapter added two new leadership approaches to the existing dichotomy of vertical and team-based leadership. These are horizontal and balanced leadership. So far, both approaches are only found in projects and can be regarded as the first project-specific leadership approaches. Through horizontal leadership, the project manager enables the use of particular skills he or she does not possess to benefit the project. Through balanced leadership, the project manager ensures that the best possible leader is appointed at any time in the project. That may answer a long-standing question in projects, which is the following.

According to the very popular Tuckman (1965) model, teams can only be performant if they go through the development stages of form, storm, norm, and perform, and with every new team member, the team starts over again from the beginning. Project settings rarely allow for this because people change frequently in projects. According to the model, these project teams can never become performant. Balanced leadership may answer why project teams can be performant even in dynamic settings with many changes in team members. That is, by ensuring the best possible leader at any point in time in the project. Hence, good leadership substitutes for the time needed to get used to working with each other on projects.

REFERENCES

Drouin, N., Müller, R., Sankaran, S., & Vaagaasar, A. L. (2018). Balancing vertical and horizontal leadership in projects: Empirical studies from Australia, Canada, Norway and Sweden. *International Journal of Managing Projects in Business*, *11*(4), 986–1006. 10.1108/IJMPB-01-2018-0002

Drouin, N., Vaagaasar, A. L., Sankaran, S., & Müller, R. (2021). Balancing leadership in projects: Role of the socio-cognitive space. *Project Leadership & Society, 2,* 1–12. 10.1016/j.plas.2021.100031

Goleman, D., Boyatzis, R., & McKee, A. (2002). *Primal Leadership: Learning to Lead with Emotional Intelligence.* Boston, MA: Harvard Business School Press.

Müller, R., Drouin, N., & Sankaran, S. (2021). *Balanced Leadership.* New York, NY: Oxford University Press.

Pearce, C. L., & Conger, J. A. (2003). *Shared Leadership* (J. L. Pearce & J. A. Conger, Eds.). Thousand Oaks, CA: SAGE Publications Inc, USA.

Thite, M. (2000). Leadership styles in information technology projects. *International Journal of Project Management, 18*(2000), 235–241.

Tuckman, B. W. (1965). Developmental sequence in small groups 1. *Psychological Bulletin, 63*(6), 384–399.

Turner, J. R., & Müller, R. (2005). The project manager's leadership style as a success factor on projects: A literature review. *Project Management Journal, 36*(2), 49–61.

3

Three Themes of Project Management

Rodney Turner
School of Civil Engineering, University of Leeds, Leeds, United Kingdom

Through the development of Project Management, we can identify three themes each giving two alternative perspectives to the management of projects. These themes are:
The systems approach versus process approach

- The focus of success being on delivering value versus the triple constraint
- Adopting a flexible approach or a strict approach to the control of projects

This reflection was stimulated by the Association of Project Management's much delayed 50th anniversary, which was eventually held at the University of Leeds in April 2003.

THE SYSTEMS APPROACH VERSUS PROCESS APPROACH

The first theme is the systems approach versus the process approach. The theory of project management was first developed in the United States, and they took a systems approach, reflected in the book by David Cleland and Bill King (1983), and the work of Harold Kerzner (2017). A project is viewed as a system, and a systems analysis approach is taken to its management. On the other hand, in Europe a project has been viewed as

DOI: 10.1201/9781003502654-3

a process (Turner, 2014, first edition 1993; Gareis and Stummer, 2008). A clearly defined process is followed to achieve successful delivery of a project. In my books, (Turner, 2014), I talk about converting vision into reality, or desires into memory. A process is followed to achieve those ends. Winch (2002) also suggests the project is a computer. Through the process we gather, store, and process information, to reduce uncertainty and to convert desire to memory, memory of what the final outcome of the project is and how we achieved it.

We can identify three main types of processes on a project:

- The investment process, which focuses on achieving project success, delivers the investment the project makes and the value and benefit it produces. Traditionally this has been called the project life-cycle, but following Voltaire, it is neither project, nor life, nor cycle. It is about the investment the project delivers, not the project, and it does not go back to the beginning, it takes us forward from vision to reality.
- The project process, where the methodology to deliver the project is applied.
- The project management process, which focuses on delivering project management success, defining, doing, and controlling the work of the project to deliver the desired project output to time and cost.

I was somewhat surprised when the Project Management Institute in their Guide to the Project Management Body of Knowledge, from the pre-edition (1987) to the sixth edition (2017), took a process approach. Their main focus was on the project management process, which they said had five elements:

1. Initiation
2. Planning
3. Executing
4. Controlling
5. Closing

The PMBoK describes what is needed to be done to deliver each of the nine body of knowledge areas through the five steps. However, in the seventh edition they have reverted to a systems approach. As we shall say below, they now focus more on delivering value than finishing on time, cost, and performance, so they focus more on the investment process. But primarily they describe a systems analysis approach to achieving value on projects.

THE FOCUS OF SUCCESS BEING ON DELIVERING VALUE VERSUS THE TRIPLE CONSTRAINT

The second theme is based on whether we view success as delivering value and benefit or finishing the work of the project in accordance with the triple constraint, of time, cost, and performance. This theme is reflected through different definitions of a project. The seventh edition of Project Management Institute's Guide to the Body of Knowledge (2021) defines a project as:

> a temporary effort to create value through a unique product, service or result.

This says the primary purpose of the project is to deliver value, and it leads the seventh edition to emphasise the investment process. The project is an investment to deliver value and benefit, and the investment process is the process that helps us to achieve that. It is about delivering project success. On the other hand, the Association for Project management in its Body of Knowledge (2019) defines a project as:

> an endeavour to deliver specific objectives subject to defined acceptance criteria, and those objectives should be delivered within constraints of time and cost.

So it is about delivering the desired project outcome to time and cost targets, achieving the triple constraint. This is project management success, and follows the project management process to plan, do and control the work to achieve that. There is a third definition due to Rodney Turner and Ralf Müller (2003). A project is:

> a temporary organization to which resources are assigned to deliver beneficial change.

The middle of this definition talks about managing resources to do the work of the project to deliver the desired output, the change, but the end talks about delivering the outcome to achieve the benefit. It does not specifically talk about finishing on time, cost, and performance, but it does mention the management of the resources, cost.

In the early days of project management, in the 1960s, the primary emphasis of project management was on delivering the desired project output to time and cost constraints. During the 1950s the emphasis was on getting the work done on time to achieve the desired objectives. Time was important, because the Americans were in an arms race. But cost suffered somewhat. In the early 1960s, as we will see again below, the new Defense Secretary, Robert McNamara, insisted on achieving the triple constraint. What is ignored is that it is usually possible to achieve only two of the three. Achieving all three is in your dreams. However, as illustrated by the seventh edition of the PMI Guide to the Body of Knowledge (2021), the emphasis is moving towards delivering value and benefit. Time and cost are important, because the project must make a profit. But it is acceptable to finish late and overspent if the project still makes a profit. People also now take much greater account of non-financial benefits. We use the example of the Gotthard Base Tunnel. Because of problems with the geology, the project was 25% late and 25% overspent. But it delivered a very worthwhile outcome. It reduced the journey time between Milan and Zurich, made for a much more comfortable ride, and enabled freight to be shifted from road to rail. The project made a profit so it was an investment success, though it was late and overspent.

ADOPTING A FLEXIBLE APPROACH OR A STRICT APPROACH TO THE CONTROL OF PROJECTS

The third theme is the difference between adopting a flexible approach and requiring a stricter approach to control. As we said above, during the 1940s and 1950s, under times of war and the arms race, the emphasis was on getting the project finished to deliver the desired output as quickly as possible. Cost suffered. Barry Klein and William Meckling (1958) wrote a now famous paper about weapons systems development. They identified two people whom they labelled the Optimist and the Skeptic. The optimist assumes they know what the outcome of the project will be and how it will be achieved. The optimist quickly closes down all other options and works towards the assumed output. Klein and Meckling say that will lead to the cheapest outcome of the if Optimists is right, but they suggest the complexity of weapons systems development means the optimist is usually somewhat wide of the mark. Change and scope creep will lead to inflating costs. The Skeptic on the other hand, assumes the precise nature of the output is not known, and so maintains several options. He does scenario planning (Drouin & Turner, 2022). As work is done, options can be closed and merged, as they work towards the actual result. If the Optimist is right, this will lead to a more expensive outcome, but will lead to less change and scope creep, and so will usually be cheaper.

When he became Secretary of Defense under President Kennedy in the early 1960s, Robert McNamara said there must be much greater control of cost. The emphasis became to follow the route of the Optimist. Define the project output as quickly as possible and freeze the design. And define how it will be achieved. This was thought to be necessary on fixed price contracts. The emphasis was on achieving time, cost, and performance, and it was what the project management professional associations preached. It was about achieving the triple constraint.

However, as suggested by Klein and Meckling, the evidence was that it didn't achieve the desired outcome. If you move quickly to freeze design, you lock yourself into expensive solutions. In the 1990s, the evidence was that if you leave uncertainty open for longer, you can achieve outcomes that are 30% to 50% cheaper. Sir Michael Latham (1994) suggested this in his report to the UK government on the construction industry. It was also the thinking behind the contract approach partnering and alliancing (Turner, 2006). By leaving the uncertainty open, the client and contractor could work together to find cheaper solutions and reduce risk. It was also the thinking behind an approach adopted in the development of oil rigs in the North Sea, called CRINE, Cost Reduction in a New Era.

So the view is that leaving the flexibility and uncertainty open for longer, rather than freezing the design as quickly as possible leads to cheaper outcomes for projects, though how this is achieved is under discussion. Klein and Meckling said it was because it avoided change and rework. Partnering and alliancing contracts, Sir Michael Latham and CRINE say it is because it allows for innovation, and you can find cheaper designs, cheaper ways of doing things, and better manage the risk. An oil field in the North Sea was estimated to cost £450 million at the end of front end design, the sanction cost was £373 million, and the final outturn cost was £290 million (Turner, 2006). The cost reductions were achieved by maintaining a flexible approach to the management of the project. Partnering and alliancing was used as the contract form, but the client and contractors were able to reduce the cost through innovation and reducing risk. If they had adopted a conventional approach to the management of the project, the sanction cost would have been £450 million, and the outturn cost may well have been higher than that with cost increases caused by scope creep, conflicts over sharing risk, and a confrontational contract management approach caused by the principal-agent governance approach (Drouin & Turner, 2002).

REFERENCES

Cleland, DI & King, WR. (1983). *Systems Analysis and Project Management*, 2nd edition. McGraw-Hill.

Drouin, N & Turner, JR. (2002). *The Elgar Advanced Introduction to Megaprojects*. Elgar.

Gareis, R & Stummer, M. (2008). *Projects & Processes*. Mainz.

Kerzner, H. (2017). *Project Management: A Systems Approach to Planning, Scheduling, and Controlling*, 12th edition. Wiley.

Klein, B & Meckling, W. (1958). Application of Operations Research to Development Decisions. *Operations Research*, 6(3), 352–363.

Latham, M. (1994). *Constructing the Team: Joint Review of Procurement and Contractual Arrangements*. Her Majesty's Stationery Office.

Murray-Webster, R & Dalcher, D. (2019). *APM Body of Knowledge*, 7th edition. Association for Project Management.

Project Management Institute. (2017). *A Guide to the Project Management Body of Knowledge (PMBOK® Guide)*, 6th edition. Project Management Institute.

Project Management Institute. (2021). *A Guide to the Project Management Body of Knowledge (PMBOK® Guide)*, 7th edition. Project Management Institute.

Turner, JR. (2006). Partnering in Projects. In DI Cleland and R Gareis (eds), *Global Project Management Handbook: Planning, Organizing and Controlling International Projects*, 2nd ed. McGraw-Hill. Chapter 20.

Turner, JR. (2014). *The Handbook of Project-based Management: Leading Strategic Change in Organizations*, 4th edition. McGraw-Hill.

Turner, JR & Müller, R. (2003). On the Nature of the Project as a Temporary Organization. *International Journal of Project Management*, 21(1), 1–8.

Winch, GM. (2002). *Managing Construction Projects*. Blackwell Science.

4

Before You Start Managing that Major Project, What You Should Know about Cost Overruns and Benefit Shortfalls[1]

Lavagnon Ika[1], Peter Love[2], and Jeffrey K. Pinto[3]
[1]Telfer School of Management, University of Ottawa, Ontario, Canada
[2]School of Civil and Mechanical Engineering, Curtin University, Perth, Australia
[3]Black School of Business, Penn State, Erie, Pennsylvania, United States

THE PROBLEM

Major projects often make the headlines not for their positive features but due to their often-publicized underperformance. Policymakers and project management practitioners thus face an uphill challenge in navigating the highly complex trials that modern major projects present. For instance, they have difficulties delivering their projects within time, budget, and specifications constraints. This problem has been referred to as the so-called 'Iron Law of Major Projects', a depressingly repetitive cycle of projects being over time, over budget, over and over again. In many cases, this short-term delivery under-performance in major projects is compounded with long-term strategic underperformance: project funders have difficulties meeting the benefits anticipated by stakeholders in the business case. This begs the question: Why do projects experience cost overruns and benefit shortfalls, and what can be done about it?

DOI: 10.1201/9781003502654-4

WHAT EXPLAINS COST OVERRUNS AND BENEFIT SHORTFALLS: ERROR VERSUS BIAS?

For practitioners and scholars alike, the reasons for such recurring project drift remain a persistent predicament. Two overarching schools of thought prevail and clash: 'the error school' and 'the bias school' (Ika et al., 2023).

The classic 'error school' associates project drift with using imperfect management techniques, making honest but erroneous forecasts or execution mistakes, lacking experience, or having inadequate data. In short, this argument suggests that projects fail because we continually do them badly. This school features several disparate contributions that suggest that projects fail to meet expectations due to scope changes, complexity, and uncertainty and underscore best practices as a way to counteract the technical and economic causes of underperformance. Evidence suggests that scope changes, for example, may account for 10 to 90% of the cost overrun figure. Notably, an audit of 20 major projects in Perth in Western Australia revealed that nearly 90% of the cost overrun occurred during the business case development period due to scope and design changes that were sanctioned by the government (Love et al., 2022).

Contrastingly, the 'bias school', which dominates contemporary thinking and practice as many governments (including in the United Kingdom, where departments now have a policy to 'debias' their project cost estimates), links project drift to a systematic distortion of logical thinking, whether intentional or not, leading to misjudgments or bad decisions, most often because of shortcuts in thinking (Flyvbjerg, 2016). Unlike the error school, the bias perspective suggests that we are predisposed, through cognitive limitations, to make snap judgments and assume plans that are simply not realistic. As the most recent and popular explanation of cost overruns and benefit shortfalls, the bias school suggests that the underlying causes of project underperformance are optimism bias (fooling oneself and not even being aware of it – delusion) and/or strategic misrepresentation (fooling others for profiteering purposes – deception), which can be curbed by ensuring decision-makers psychological and political accountability. Empirical evidence suggests that optimism bias, for example, may affect 60% of projects and reduce project performance up to 20% (Love et al., 2022).

WHAT EXPLAINS COST OVERRUNS AND BENEFIT SHORTFALLS: HIDING HAND VERSUS PLANNING FALLACY?

Two basic arguments power these competing schools of thought: the Hiding Hand Principle and the Planning Fallacy Principle. So, while the error school may draw on Hirschman's (1967) Hiding Hand Principle as an exemplar theory, the bias school employs the theory of the Planning Fallacy Principle proposed by Kahneman (2011). As empirical evidence demonstrates, the former theory suggests ignorance can be good as it can lead to 'stumbling into success', owing to unforeseen creativity on the part of planners and managers, while the latter theory suggests ignorance is always bad as it leads to starting projects that should never have started. As these two competing theories offer important implications for understanding cost overruns and benefit shortfalls, the error *versus* bias debate has been termed the Planning Fallacy debate (Ika et al., 2022).

The following two major projects with contrasted fates, 47 Rōnin – a 'flop' – and the Hoosac Tunnel – a 'near miss' – provide fitting illustrations of both schools of thought and the Planning Fallacy debate.

47 Rōnin: The Bias School and the Planning Fallacy Theory

In 2008, *47 Rōnin*, an American martial arts film project, was proposed by Universal executives. Over the course of its development, the project experienced many script and scope changes. After two postponed releases, the $175 million mega-production was finally released on Christmas 2013, more than one year later than originally expected, with a final cost of over $225 million. Ultimately, *47 Rōnin* finished sixth on opening weekend and grossed only $20 million in its first five days at the US box office. The film became one of Hollywood's costliest box office busts of 2013.

Looking back, the *47 Rōnin* project suffered from optimism bias – the belief that decision-makers are less likely to face risks than statistical reality warrants – on the part of its promoters, leading them to underestimate the project's times, costs, and risks, and overestimate its benefits and likelihood of success. Consequently, the project should never have been started, at least according to some experts. Flyvbjerg (2016) notes, 'the average project is in fact undermined by a double whammy of substantial cost hikes compounded by substantial benefits shortfalls' (p. 176). This kind of project disaster is a good example of what Kahneman (2011) calls the *Planning Fallacy*.

The (Malevolent) Planning Fallacy, citing evidence of the tendency for projects to over-promise and under-deliver, suggests that forecasts of project schedules, costs, and benefits are by and large unrealistically close to best-case scenarios. According to the proponents of the Planning Fallacy, there are systematic biases (including biases in decision-making) in how projects are selected, managed, and reported on, which explain a project's underperformance. So impactful has been the bias school that in 2016, Virginia Raggi, then-mayor of Rome, withdrew her city's bid to host the 2024 Olympics, citing historical evidence of huge cost overruns from other cities' Olympic experiences.

The Hoosac Tunnel: The Error School and the Hiding Hand Theory

In 1819, the Hoosac Tunnel was envisioned as part of a canal project to connect the cities of Boston and Albany, but the Hoosac Mountain was standing in the way. Construction finally began in 1851 on the Hoosac Tunnel. The project, however, took 24 years to complete, had a final cost of more than ten times the initial budget, which was forecasted at $2 million, and experienced unprecedented challenges, including several tunneling device failures. In spite of a steady string of setbacks, the engineers still discovered unexpected ways to deliver the tunnel.

Looking back, if Crocker and colleagues had known all the difficulties they would confront in advance, they would most likely never have undertaken this project. This would have been a great loss because the longest tunnel in North America then became one of the greatest engineering feats of the 19th century. In addition, it successfully provided a much-needed commercial link between the state of Massachusetts and the West, which later on was critical to trade and economic development. As Ika (2018) cogently notes: 'This is an example of how ignorance and poor planning can make decision-makers low-ball the real costs and challenges of a project, and heavily understate their own creativity to overcome project-related problems' (p. 371). This failure-success paradox, 'near miss,' or silver lining, illustrates how ignorance needs not be an automatic death sentence for projects; an idea Hirschman (1967) proposed in what he coins the *Principle of the Hiding Hand*.

The (Benevolent) Hiding Hand, a key theory that is drawn upon by the error school protagonists, suggests that this propensity for planners and managers to overestimate the benefits and underestimate the costs and difficulties of their assigned projects is not always bad, as creativity may come to the rescue in unforeseen circumstances. In other words, the Hiding Hand proponents suggest that data arguing that most projects are 'dead on arrival' due to systematic errors in initial estimation coupled with wilfull mis-representation, as the Planning Fallacy claims, grossly over-represents this position. They argue that such premeditated culpability is simply one potential source of project failure but does not consider how initial miscues may lead to redirection, recovery, and ultimate success. Creativity may hence, in some cases, come to the rescue and make projects ultimately 'stumble into success'. Indeed, many UNESCO World Heritage sites, such as the Sydney Opera House in Australia, would not have been undertaken had their promoters known the actual costs and difficulties they would encounter.

IT'S NOT ERROR OR BIAS BUT BOTH!

Does bias (e.g., the Planning Fallacy or the tendency for projects to over-promise and under-deliver) trump error (e.g., scope changes, complexity, and uncertainty) as the root cause of project underperformance? Both bias and error may be at play in many projects.

THE ERROR AND BIAS SCHOOL AND THE THEORY OF THE SAVVY HAND

The above examples of a flop and a near-miss suggest that the Planning Fallacy and the Hiding Hand may help tackle the question of what forces best explain cost overruns and benefit shortfalls. In particular, the Planning Fallacy predominates when promoters practice profiteering and corruption as they seek to profit from information asymmetries and other market and governance failures. In contrast, the Hiding Hand fits well with organizational contexts where risk-taking and problem-solving behavior prevail (Ika et al., 2022).

Both the Planning Fallacy and the Hiding Hand fail to account for the shades of grey between success and failure in projects. Indeed, the Hiding

Hand concerns projects that are failures regarding cost, for example, and successes regarding benefits delivery. The Planning Fallacy manifests in projects that are seen as all-around failures, that is delivered under budget and under benefits. Because neither theory sufficiently describes the current state of project underperformance, two additional causes have been added for a total of four reasons for persistent problems (Anheier, 2017), including the Hiding Hand or Benevolent Hand, the Planning Fallacy or Malevolent Hand, the Passive Hand, which leads decision-makers to stifle creativity and avoid risks, and the Protecting Hand or precautionary principle which tackles ignorance through risk management and scenario-planning. While these four reasons collectively offer superior explanation for project underperformance, they do not shed light on the shades of grey between success and failure or between optimism and pessimism.

Consequently, we propose the theory of the Fifth Hand, which suggests a pluralistic approach to understanding major project underperformance by combining elements and assumptions embedded both in the Planning Fallacy and the Hiding Hand and recognizing that effective diagnosis of underperformance requires a willingness to accept the existence of multiple causes, both bias and error. For convenience's sake, we call the Fifth Hand 'Savvy Hand' in the rest of this chapter as it summons practitioners to be project 'savvy' (see Gigerenzer, 2014 about risk 'savvy'). The Savvy Hand does not solely focus on cost overruns and unforeseen benefit success like the Hiding Hand or merely emphasizes cost overruns and benefit shortfalls like the Planning Fallacy. The Savvy Hand would provide project planners, managers, and teams with an understanding of when explanations involving the Planning Fallacy or the Hiding Hand may well work. As such, the Savvy Hand seeks to provide an understanding of the circumstances in which projects work or not (Ika et al., 2022).

THE VA AURORA HOSPITAL: THE ERROR AND BIAS SCHOOL AND THE SAVVY HAND THEORY AT WORK

The Aurora Veteran Affairs (VA) Hospital, Colorado (USA), is a good illustration (Ika et al., 2023). The Aurora VA Hospital project's concept goes back to 1995. Continuing in fits and starts, it has evolved from a shared hospital facility to a stand-alone VA medical campus. The Aurora VA Hospital's project aimed to create a state-of-the-art facility to treat about 400,000 veterans suffering from various disabilities. Initially budgeted for

$328 million in 2004 and re-pegged at $678 million in 2011, the as-yet uncompleted project costs ballooned to a new price tag of $1.7 billion by 2015, notwithstanding scope changes to cut costs, including the abandonment of the planned Post Traumatic Stress Disorder treatment center and nursing home. Ultimately, the US Congress handed the responsibility to complete the project to the Army Corps of Engineers. The massive US$2 billion, 11-building, 1.2 million-square-foot, 31-acre medical center advanced far enough toward completion to support a formal ribbon-cutting ceremony in 2018 (a decade behind schedule), while millions in claims are still being adjudicated amidst a flood of litigation between the VA Aurora Hospital and their primary construction contractors. Three years after completion, the VA Aurora Hospital became one of the most expensive health facilities in the world. Additional expenses had to be incurred between 2018 and 2021 to address a series of malfunctioning, poorly installed, or missing elements. Ultimately, the Aurora VA Hospital had to spend an additional US$785 million to outfit the complex and resolve the problems encountered.

The controversial history of the development of the VA Hospital in Aurora illustrates the critical nature of creating a means for satisfying the bias *versus* error debate. Indeed, some observers, such as the Congressional Budget Office, argue that this is a textbook case of over-optimism coupled with deliberate concealment of costs; others suggest the true culprit was the snowballing of errors, misjudgments, and poor communication and collaboration among critical stakeholders. In addition, scope changes occurred due to Congress insisting on after-the-fact cost cutting, as with the US$20 m PTSD building. The implications here are important because an improper diagnosis of fundamental problems can lead to serious misapplication of corrective steps or 'fixes' on the part of project organizations attempting to right a rapidly deteriorating status.

PRACTICAL RECOMMENDATIONS

Reducing error and/or bias is at the core of the three schools of thought. The error school suggests the need to apply best practices to counteract slippages due to technical and economic causes, including improving risk analysis, cost forecasting, and project implementation. However, these recommendations fall short, as we have seen in the case of the Boston 'Big Dig' where it has been conceded by the risk manager of the project that error, and in particular the poor management of the project's complex integration, was the major cause of its significant cost overruns (Ika et al., 2023).

Arguing that historical data demonstrates the problem is NOT cost overruns and benefit shortfalls but instead cost *underestimation* and benefit *overestimation*, the bias school contends that subsequent post-analysis 'smoothing', or correction mechanisms are required to develop 'true' estimates. So resonant are the warnings of Planning Fallacy champions that public sector decision-making has been significantly affected. For example, the requirement by the Irish government that all new project proposals apply Reference Class Forecasting (RCF) to debias cost estimates is instructive. Put simply, RCF is a risk management tool that helps determine an 'uplift' percentage that should be added to the initial cost estimates of projects under consideration, based on data drawn from previous and similar projects' results (the 'reference class'). The assumption is that a database of hundreds of past similar projects should demonstrate an average budget overrun, which is then simply added to the project's initial cost estimate to approximate the 'real' likely cost of the proposed project. While this idea sounds good in theory, the record of RCF to actually demonstrate true costs is very much open to debate, as several famous projects (such as London Crossrail) employed RCF and still experienced billions of dollars in cost overruns.

Combining the error and bias schools still only leads to a sterile debate between the Cassandras – the overly pessimistic – who hold a bias for despair and think that many projects should not have been started, and the Pollyannas – the excessively optimistic – who prefer clinging to hope and firmly believe that many projects would in the end 'stumble into success' (Ika et al., 2023). What is needed is a third character in the world of projects: the Januses (named for the Roman two-headed god) – who are half the optimistic Pollyannas and half the pessimistic Cassandras and appreciate the shades of grey between success and failure in projects as well as the tensions they cause. Thus, the error and bias school invites practitioners not only to combine best practices to deal with scope changes, complexity, and uncertainty with budget uplifts to ensure the accountability of decision-makers.

For example, recognizing that both 'behavioral biases' and 'organizational errors' may thwart project success, practitioners may have to apply at different phases of the project lifecycle, including forecasting, organizing, and executing, optimism bias uplifts, aligning incentives between planners and bidders as well as principals and agents, put in place a 'double-blind peer-review' of project proposals, tackle deficient project delivery capacity, and opt for modular design and fast implementation of major infrastructure projects (Lovallo et al., 2023).

In addition, practitioners may also consider the rational (e.g., plans and biases as they pertain to the project mandate or specific objective), political (e.g., power of the principals outside the core project team) and psychosocial (e.g., power of the agents inside the core project team) aspects of project delivery, and explore the contribution of heuristics or rules of thumb such as 'Your biggest risk may be you but your biggest asset is also you' or 'Plan your work, work your plan but be ready for welcome and unwelcome surprises down the road' in the face of uncertainty (Ika and Saint-Macary, 2023).

Table 4.1 summarizes the key insights from this chapter.

TABLE 4.1

What Explains Cost Overruns and Benefit Shortfalls, and What can be Done About It

Schools Characteristics	Error	Bias	Error and Bias
Key Theory	Hiding Hand	Planning Fallacy	Savvy Hand
Key Causes	Scope changes, complexity, and uncertainty	Behavioral biases including optimism bias (or delusion) and strategic behaviors such as strategic misrepresentation (or deception)	Scope changes, complexity, uncertainty, behavioral biases, and strategic behaviors
Key Solutions	Best practices for cost forecasting, risk analysis, and project execution	Debiasing approaches such as budget 'uplifts'	Combine best practices (including heuristics) and debiasing approaches
Key Aim	Counteract technical and economic causes of project drift	Lift the veil on psychological and political causes of project drift to ensure accountability of decision-makers	Deal with all causes of project drift
Key Character	Pollyannas or over-optimists	Cassandras or over-pessimists	Januses or half optimists and half pessimists

NOTE

1 This chapter has been presented as a part of the University of South Africa (UNISA) thought leadership engagement series in 2022 and at the Swedish Transport Administration Workshop in 2023. Chunks of the chapter have been adapted from Ika et al. (2023).

REFERENCES

Anheier, H. K. (2017). Infrastructure and the principle of the hiding hand, in Wegrich, K., Kostka, G., Hammerschmid, G. (Eds). *The Governance of Infrastructure*. Oxford University Scholarship Online.

Flyvbjerg, B. (2016). The fallacy of beneficial ignorance: A test of Hirschman's hiding hand. *World Development*, 84, 176–189.

Gigerenzer, G. (2014). *Risk savvy: How to make good decisions*. Penguin Group.

Hirschman, A. O. (1967). *Development projects observed*. The Brookings Institution.

Ika, L. A. (2018). Beneficial or detrimental ignorance: The straw man fallacy of Flyvbjerg's test of Hirschman's hiding hand. *World Development*, 103, 369–382.

Ika, L. A., Love, P. E. D., and Pinto, J. K. (2022). Moving beyond the planning fallacy: The emergence of a new principle of project behavior. *IEEE Transactions on Engineering Management*, 69(6), 3310–3325.

Ika, L. A., Pinto, J. K., Love, P. E. D., and Paché, G. (2023). Bias versus error: Why projects fall short. *Journal of Business Strategy*, 44(2), 67–75.

Ika, L., and Saint-Macary, J. (2023). *Managing fuzzy projects in 3D: A proven, multi-faceted blueprint for overseeing complex projects*. McGraw-Hill Education.

Kahneman, D. (2011). *Thinking fast and slow*. Doubleday Canada.

Lovallo, D., Cristofaro, M., and Flyvbjerg, B. (2023). Governing large projects: A three-stage process to get it right. *Academy of Management Perspectives*. 10.5465/amp.2021.0129

Love, P. E. D, Ika, L. A., and Sing, M. C. (2022). Does the planning fallacy prevail in social infrastructure projects? Empirical evidence and competing explanations. *IEEE Transactions on Engineering Management*, 69(6), 2588–2602.

5

Why "Well-Padded" Projects Continue to Be Late[1]

Jeffrey K. Pinto[1] and Kate Davis[2]
[1]Black School of Business, Penn State, Erie, Pennsylvania, United States
[2]Cranfield School of Management, Cranfield, United Kingdom

INTRODUCTION

When organizational members create activity duration estimates that afford them maximum protection (padding to get their work done with a margin of safety), the law of unintended consequences invariably comes into play. Many of these mistakes are a result of the creative and often self-serving means to create sufficient slack time so they can feel confident in delivering their activities on time. Based on these behaviors, a fair question could be asked: with so much artificial slack time embedded in project schedules, how is it that a project could ever be late? Surely the combined weight of individual estimates, team supervisor padding, and anticipating top management cuts would ensure that there was plenty of time in the project plan to ensure that all activities would finish comfortably within their estimates. And yet, common observation and personal experience suggest that in spite of ample padding, projects still over-run their schedules, leading to the obvious question: why? As we will establish in this chapter, the short answer to this question is that while humans are wonderfully creative at finding ways to add padding to their activity estimates, we are equally adept at squandering all this newly-created slack once the project gets underway. This chapter will address the flip side of the coin: why is it that our well-padded project schedules continue to run late?

As we consider the reasons why a project that received injections of extra safety can still go off the rails and run weeks, months, or even years late, it

DOI: 10.1201/9781003502654-5

is first important to note that not all of the reasons to be examined are the result of human error—unintentional or otherwise. That is, we will see that in some cases, other conditions outside of managerial control can result in wasting huge amounts of carefully gathered activity safety. It is important, then, to carefully examine each of these reasons why projects continue to run late and apply an internal checklist to this set: which of these behaviors are we guilty of? What can we reasonably control and what actions, while outside of our direct oversight, can still be corrected if we recognize them and apply corrective responses. Finally, it is also important to note that not every reader will recognize each of these dynamics in their own organization. We have compiled a set of behaviors from literature (our own work and studies of other scholars), anecdotes from students, and personal experience over many years. Some may find that their organization is only guilty of a small sub-set of these behaviors, while others will recognize the long litany of responses here as all having a role in the sub-performance of their projects.

REASON 1: THE LEARNING CURVE MODEL (OR, WHAT HAPPENS WHEN GOOD INTENTIONS MEET REALITY)

Speak with project team members in most organizations and you will hear versions of a similar story: they are extremely busy, responsible for work on multiple projects at the same time, and feeling overextended and stressed out. Project team members are a valuable and often overworked group of people who are trying to stay afloat in a sea of competing demands for their time and attention. With these multiple and competing pressures, it is natural to resort to juggling: trying to keep multiple balls in the air at the same time while "nibbling"; offering incremental attention to each of our commitments incrementally. We work on Project A for a time, then shift to Project B, before moving on to another commitment, in a continuous cycle.

This multitasking behavior has another implication, what has been called the "term paper model" (Graham, 1989) and should be readily recognizable to most of us. When assigned a project—such as a term paper while we were undergraduates—there was an implicit connection between the time we had to complete the assignment and the amount of work necessary to get it done. Figure 5.1 shows this model.

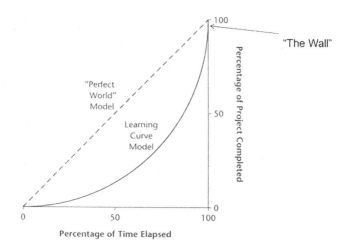

FIGURE 5.1
Term paper model for project activity completion.

The "perfect world" response to receiving an assignment, like a project activity or the college term paper, would be to track a direct linear relationship between time and percentage of the task completed; e.g., after 50% of the time has elapsed, we would expect to have completed 50% of the task and so forth. This is often not the case and reality reflects a "Learning Curve Model" when new project tasks are assigned.

This is either due to other, more immediate commitments they wish to complete, or the need to develop a level of expertise and knowledge (the learning curve) before they actually start the task. This creates a noticeable lag effect at work going around in circles, while little (or no) work is being done on the task. We note that this lag may not be the result of disinterest or a lack of willingness of the worker to do the assignment; it is usually more often associated with people trying to multitask a number of simultaneous commitments, with some due dates looming earlier than others (Pinto, 1999).

Just as our undergraduate experience with semester-long term paper assignments demonstrated, it is difficult to generate motivation to steadily work on a task whose completion date is off in the future when faced with more immediate concerns, prompting us to put off the start for as long as possible. In fact, Graham (1989) found that it is not uncommon for project team members to delay making meaningful progress on their activities for extended periods of time, mirroring the learning curve model, until the realization that significant time has elapsed (perhaps 50%), while nothing of note had been done. As time continues to unwind and the due date, or "The Wall," looms ever closer, a degree of nervous energy—even panic—sets in and work proceeds at a much more rapid pace. However, there is often a companion pressure that all project managers have faced at some point in their careers: the urgent request of project team members for more time to complete the task. In effect, the fear of running out of time (hitting "The Wall") becomes a very real possibility, given the amount of time that has transpired and their current low level of performance. Project team members who have delayed working on the task until its pending delivery date is looming in the immediate near term will approach their project manager, convinced that they cannot possibly complete the assignment in the time available, and ask for an extension. Many novice project managers are viewing the same data and feeling their own sense of concern about this lack of activity (and project) progress and the need to develop plausible remedial actions to address these potential delays. Ironically, when considering what to do in this situation, if we are not careful, we can actually make a bad situation much, much worse.

When faced with pending activity delays, there are two common solutions identified by the project manager: 1) adding resources and 2) employing overtime (pushing their team to work harder). Both of these solutions seem attractive, but actually are almost always counter-productive, as the implications of these mistakes become apparent. Let's consider both of these "solutions" in turn, in order to understand the initial motivation for using them and the subsequent problems they often generate.

1. **Wrong Solution One: Adding Resources to the Late Activities**
   Suppose we are facing a situation illustrated by the Learning Curve Model (see Figure 5.2), in which an ongoing critical activity is in danger of becoming late. The pressure on the project manager to "fix it" in this case can become overwhelming, forcing some critical choices, the most obvious of which is to add more people to the task under the assumption that if five people cannot get it done on time, eight people should certainly be able to finish the assignment (Brooks, 1975). Adding resources to late activities is an alluring idea, as it implies that the project manager sees the problem and is doing everything—in real time—possible to address it. Unfortunately, it typically creates a huge unforeseen problem, most often referred to as "Brooks' Law."

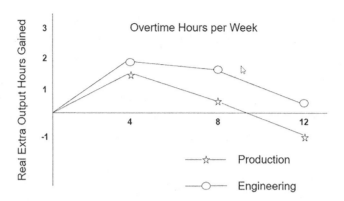

FIGURE 5.2
Impact of overtime on project activity productivity.

(**Cooper, 1994**).

Brooks' Law suggests adding resources to late project activities only makes them later. To illustrate Brooks' argument, imagine a project activity for which you are responsible and to which you have assigned five team members. Suppose that the delivery of the activity is lagging, getting further and further behind the target delivery date, leading to some panic on the part of the team. In your desire to help, you assign two additional resources to the activity. The result, however, is only to delay it further because your "remedy" has effectively just taken two of the five workers out of action, as they are pulled away for the activity to bring the two new hires up to speed. Now, the project is even further delayed, and we panic even more, perhaps assigning two *additional* workers to help fix the problem. Again, the new net effect is to pull two more people off the task for learning curve training, leaving us with only one person actually working on the task! A project that was running late has now fallen critically behind schedule, all due to the desire to fix the problem.

Brooks' argument is that it takes time for people added to a project to become productive and an ongoing, but late activity is about the worst starting point for bringing these new team members up to speed. While Brooks' Law is a bit of an over-simplification, his perspective has been shown time and again to occur in multiple project settings and it offers a powerful way of viewing project schedules in light of activity learning curves and what we think are logical remedial solutions but are actually counter-productive.

2. **Wrong Solution Two: Working Overtime**

   A second common response to late activities that, while often employed, is actually equally counter-productive is the use of overtime to get back on track. The work of author and consultant Ken Cooper on the topic of project rework and schedule delays is particularly important in understanding the fallacy of using overtime to recover lost time on our project. In his studies of hundreds of projects over many years, he found that the use of overtime quickly degraded overall project quality (see Figure 5.2). Examining project team members from two disciplines—engineering and production—Cooper (1994) charted their real extra output gained relative to the number of overtime hours worked each week.

   The results are striking: four additional hours of overtime a week for both groups resulted in less than two hours of extra output and the more overtime worked, the more rapidly effective output dropped, until 12 hours of overtime per week led to *near-zero or even negative* consequences. Cooper found clear evidence that working longer hours led to a general drop in quality of work, resulting in the need for additional rework to be done to correct the errors made by fatigued team members. Causes of rework have been studied extensively in the project management literature and in some cases, the costs of rework can become enormous, with some classes of project (IT and construction) contributing over 50% of a total cost overrun incurred and can increase schedule overrun by 22% (Love and Sohal, 2003). In the face of damning data such as these, it is important to recognize that seemingly harmless efforts like overtime intended to get a project back on track can serve as a backdoor into large cost increases through resulting rework by tired employees.

REASON 2: MERGING PROJECT PATHS

We noted at the start of this chapter that it is important to distinguish between willful (even if inadvertent) behaviors that can remove the project safety we embedded in our original plan and other events or conditions that have the same result, merely as a function of the way we constructed the project plan. The influence of merging paths is an excellent example of this situation (Leach, 1999). When a project schedule includes merge points, either in the form of individual activity merge points or the merging of longer project paths within the network, we can see the potential for significant delays, even if they are not our direct fault. Consider a situation in which our project has three parallel paths that come to a merge point: Path A is 15 days late, Path B is right on time, and Path C is actually 15 days early. Good news, right? Well, actually, no. The status of the overall project is constrained by the slowest merging path, which means that all the hard work of the resources responsible for Paths B and C are lost due to the tardiness of Path A. Because these paths merge prior to the start of the next activity, the project is now 15 days late moving forward. In this way, a project schedule falls hostage to the slowest performer, failing to account for the potential delays of one path by punishing those working on all the paths in the project network.

REASON 3: POSITIVE VARIANCE IS NOT PASSED ALONG

Positive variance is defined as finishing early. When an activity is estimated to take ten days and instead, we are able to finish it in eight days, we have generated two days of positive variance. The question now is simple: what do we do with those two extra days? Do we present our boss with the completed activity two days early? Perhaps. On the other hand, there may be valid reasons to hang on to that assignment, waiting until the originally agreed upon due date before submitting our work. Remember that if there is an atmosphere of distrust between the worker and the project manager (or top management just waiting for the excuse to take a knife to future project schedules), the act of submitting an assignment early can actually trigger a negative reaction; i.e., "You told me you needed 10 days. Why was your estimate 20% off?" Another ironic and unintended consequence could be to foster deeper distrust between the worker and manager when asked to produce estimates for future assignments. "Last time you told me it would take 10 days and you got it done earlier. This time, I expect you to finish your work in seven days. No excuses!" Thus, a well-meaning gesture demonstrating high productivity can actually boomerang and lead to greater distrust and harder future standards. When project team members face a culture of distrust, they are far less likely to make themselves part of the solution for speeding up a project, including by sitting on positive variance (Leach, 2014).

REASON 4: THE EFFECTS OF MULTITASKING

Earlier in this chapter, we discussed some of the implications of multitasking on project activity completion and the term paper model. Here, we would like to address the idea directly, applying implications first articulated by Goldratt (2017) in his classic book, Critical Chain. Goldratt noted that the greater the degree with which project resources are subject to competing demands that result in multi-tasking, the easier it becomes for individual activity responsibilities to slide. He illustrated this idea in Figure 5.3, where he demonstrated a simple complication to a relatively straightforward resource assignment.

Suppose members of our project team were permitted to schedule activities in a manner that allows them to address these activities in order, completing Activity A before starting Activity B and finally, addressing Activity C. Under this circumstance, each activity would be completed in ten calendar days. However, when resources are

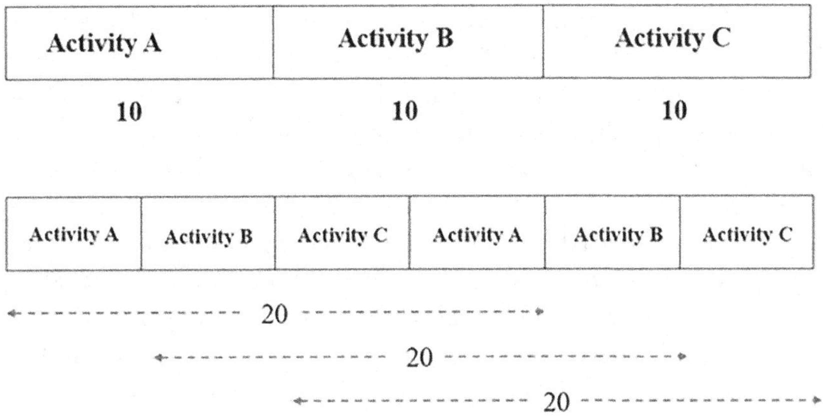

Activity A	Activity B	Activity C
10	10	10

Activity A	Activity B	Activity C	Activity A	Activity B	Activity C

← - - - - - - - - - - - 20 - - - - - - - - - - - →

← - - - - - - - - - - - 20 - - - - - - - - - - - →

← - - - - - - - - - - - 20 - - - - - - - - - - - →

FIGURE 5.3
The effects of multi-tasking.

(Goldratt, 2017).

multitasking, they are trying to work a portion of their time on each activity without being able to fully complete any one of them. So, in the case where Sally can only work five days on Activity A before feeling the pressure to switch her attention to Activity B, and then after an additional five days move on to Activity C, we can see the practical implications of this multitasking, as it lengthens the completion time for each of these activities to 20 days duration, through no other reason than multitasking pressure. It has also been pointed out that this is actually a simplistic model in that it assumes a seamless transition from one activity into another with no loss of productivity. Suppose we also factored in the time needed for such transitions, to reacquaint ourselves with the work we were doing and catch back up through the learning curve. Each transition could take an additional day or more, which effectively pushes out the activity completion dates even further. The key point is that the more we expect our project team members to multitask without taking these additional pressures into our project schedule, the more we set ourselves and our projects up for embedded and indeed, expected, delays (Dye, 2006). The problems inherent in multitasking, or work fragmentation, have long been recognized by behavioral scientists, education theorists, and neurologists.

One implication of workplace fragmentation involves the potential disconnect between various stages of a project; as the work becomes increasingly incremental and difficult to track, it is harder to coordinate project stages or activities across the project life cycle (Alashwal and Fong, 2015). Moreover, we know that in spite of our best intentions, worker productivity falls under conditions of multitasking (Rosen, 2008) and general job dissatisfaction increases. Further, in line with our contention regarding the cost of changing tasks mid-stream, evidence shows that switching attention between different tasks results in a 50% longer time to finish those tasks, compared to focusing on one task through to completion before starting the next one (Gendreau, 2007).

To be clear, the above is not an argument to abolish multitasking in all circumstances, as many organizations are resource-constrained and are forced to attempt a number of activities with limited resources. We do need to be clear, however, that our multitasking assignments, logical or defensible though they may be in our eyes, carry significant negative consequences for projects. The more organizations rely on policies that promote excessive multitasking among their project team members, the greater the likelihood they will encounter negative effects on productivity, morale, and a sense of personal efficacy.

REASON 5: PARKINSON'S LAW

Parkinson's Law suggests that **work expands to fill the time available for its completion** (Parkinson, 1958). In project management, the practical effects of Parkinson's Law are particularly apparent in the loss of project slack time, as team members and their managers first work to create project slack, but then are prone to wasting it through time mismanagement. Putting this another way, it has been observed that a "loose deadline" leads to a decline in worker performance and to delays in their activity completion (Gutierrez and Kouvelis, 1991). As a result, project organizations have to deal with the twin challenges of inauthentic project schedules, coupled with the realization by some team members that, because the activity contains plenty of slack, there is an implicit assumption that it is not necessary to push hard to finish early; i.e., some time before the activity's late finish. In battling the effects of Parkinson's Law, organizations have found creative methods to help workers motivate themselves to seek early completions, including the creation of incentive-based systems that reward workers for early finish (Chen and Hall, 2021).

You will recall that we developed this chapter as an answer to a thorny question: with the potential for so much extraneous padding (slack) injected into the project schedule, how could it ever be the case that a project will finish late? This chapter attempts to offer several reasons why all the extra slack, so assiduously acquired by members of the project team and their supervisors, can be wasted through a series of practices, behaviors, and attitudes on display during project execution. In other words, we acquire lots of slack just to turn around and waste it in creative and not-so-creative ways. The net result for our projects, if we are not carefully monitoring our propensity toward these behaviors, is to make critical mistakes in both the creation of schedules, as well as the actions taken during project development. Planning that creates unnecessary slack and actions that waste time are two sides of the same coin: a recognition that for all the science and theory we have created to support our projects, they still remain at the mercy of behavioral miscues that may sabotage a team's and project manager's success.

NOTE

1 Portions of this chapter were adapted from: Pinto, J.K. (2022), No project should ever finish late (and why yours probably will, anyway), *IEEE Engineering Management Review*, 50, 181–192.

REFERENCES

Alashwal, A. M. & Fong, P. S.-W. (2015). Empirical study to determine fragmentation of construction projects. *Journal of Construction Engineering Management*, 141(7), 04015016.

Brooks, F. P. (1975). *The Mythical Man-Month: Essays on Software Engineering*. Reading, MA: Addison-Wesley.

Chen, B. & Hall, N. G. (2021). Incentive schemes for resolving Parkinson's law in project management. *European Journal of Operational Research*, 288(2), 666–681.

Cooper, K. G. (1994). The $2000 hour: How managers influence project performance through the rework cycle. *Project Management Journal*, 25(1), 11–24.

Dye, L. (2006). Managing multiple projects: Balancing time, resources, and objectives. In Dinsmore, P. C and Cabanis-Brown, J. (Eds.), *The AMA Handbook of Project Management, 2nd Ed.*, pp. 333–342. New York, NY: AMACOM.

Gendreau, R. (2007). The new techno culture in the workplace and at home. *Journal of American Academy of Business*, 11(2), 191–196.

Goldratt, E. M. (2017). *Critical Chain*. New York, NY: Routledge.

Graham, R. (1994). Personal communication.

Graham, R. (1989). *Project Management as if People Mattered*. Bala Cynwyd, PA: Primavera Press.

Gutierrez, G. J. & Kouvelis, P. (1991). Parkinson's law and its implications for project management. *Management Science*, 37(8), 990–1001.

Leach, L. P. (2014). *Critical Chain Project Management*. Norwood, MA: Artech House.

Leach, L. P. (1999). Critical chain project management improves project performance. *Project Management Journal*, 30(2), 39–51.

Love, P. E. D. & Sohal, A. S. (2003). Capturing rework costs in projects. *Managerial Auditing Journal*, 18, 329–339.

Parkinson, C. N. (1958). *Parkinson's Law: The Pursuit of Progress*. John Murray: London.

Pinto, J. K. (1999). Some constraints on the theory of constraint: Taking a critical look at the critical chain. *PMNetwork*, 13, 49–51.

Rosen, C. (2008). The myth of multitasking. *New Atlantis*, 20, 105–110.

6

Are Agile Projects More Successful?[1]

Pedro M. Serrador
Northeastern University, Toronto, Ontario, Canada

"It is a bad plan that admits of no modification."

Publilius Syrus (~100 BC)

Projects continue to proliferate in society today, in both the public and private sectors of the economy. Investments in projects number in the trillions of dollars annually. Just as ubiquitous as these projects, unfortunately, are their significant failure rates. The CHAOS reports have identified the current state of project success rates across organizations, noting that in spite of the much higher visibility and importance placed on project performance, failure rates have remained high and relatively stable across over a decade of research (The Standish Group, 2011). Further, specific examples of project failures shed light on the impact they have on organizations. Consider, for example, the following:

- Joe Harley, then-CIO at the Department of Work and Pensions for the UK government, stated that only 30% of technology-based projects and programs are a success – at a time when taxes are funding an annual budget of £14 billion (about US$22 billion) on public sector IT, equivalent to building 7,000 new primary schools or 75 hospitals a year (Ritter, 2007).
- "Motorola's multibillion-dollar Iridium project … could be considered a success on the basis it was 'on time' and 'on budget' from

an engineering point of view, but was a catastrophic commercial failure because it did not adjust to what was being learned about the changing business environment" (Collyer, Warren, Hemsley, & Stevens, 2010, p. 358). The project team and management at Motorola failed to see that during the course of the project, quickly expanding cell phone networks would undercut Iridium's satellite phone business model.

It is with this setting in mind that researchers and practitioners began seeking alternative methods for project implementation, recognizing that traditional models for planning and execution may not be optimal or tuned for the specific challenges that projects face. Indeed, it is due to these challenges that "light weight" project management techniques such as Agile have been gaining popularity since first developed (Dybå & Dingsøyr, 2008).

Part of the ethos of Agile methods is that less initial planning is better and an evolutionary process is more efficient (Dybå & Dingsøyr, 2008). Agile methodologies contrast with traditional project management approaches (such as waterfall) by emphasizing continuous design, flexible scope, freezing design features as late as possible, embracing uncertainty and customer interaction, and a modified project team organization. Further, Agile is described as iterative and incremental, seeking to avoid the standard approaches that emphasize early design and specification freeze, a fixed project scope, and low customer interaction.

These more traditional project development approaches pursued a goal of logical sequencing that required deliverables to be set in advance and project development evaluated based on performance at a series of capabilities gated reviews. Unfortunately, evidence continues to accumulate suggesting that a rigid development process can result in significant downstream pathologies, including excessive rework, lack of flexibility, customer dissatisfaction, and the potential for a project to be fully developed, only to discover that technological advances have eclipsed the need for it. So, for example, to revisit the post-mortem analysis of Motorola's Iridium project, it became clear that in dynamic environments, projects need to cope with changes in technology during the course of their development both for technology and other projects. If assumptions fail, unsuccessful projects can often result. "While useful as a guide, excessive detail in the early stages of a project may be problematic and misleading in a dynamic environment" (Collyer, Warren, Hemsley, & Stevens, 2010, p. 109).

Though Agile methods are continuing to gain in popularity and are spreading beyond their original birthplace among software development projects (Dybå & Dingsøyr, 2008), little research has been done as to whether Agile project truly are more successful. The majority of research examining its usefulness has been anecdotal, small case studies, or research limited by sample size, industry, or geography.

We did a large-scale quantitative study, looking for evidence that Agile methods work better than traditional approaches for achieving project success (Serrador & Pinto, 2013). Agile has become widely used and a generally accepted approach for planning and executing projects in IT settings. There is a wealth of anecdotal and case-study information pointing to the utility of the Agile process; however, a detailed study was needed testing the efficacy of the Agile philosophy as it directly relates to project success. This chapter reports on the results of a recently completed study of projects and their success rate. We investigated the efficacy of Agile on different dimensions of project success, across multiple industries, in order to identify the degree to which Agile can be directly linked to project success, its viability across multiple project environments, and the potential for intervening (moderator) variables to affect this relationship.

As early as 1958, Koontz noted that "no effective manager makes a plan and then proceeds to put it into effect regardless of what events occur" (Koontz, 1958, p. 54). (Hällgren, 2005, p. 18) note the inevitability of deviation in project plans, suggesting the solution lies not in more sophisticated initial plans but in methodologies that can facilitate actions to resolve deviations. In the IT project environment, this need for improving the planning process has increasingly led companies away from the traditional, front-end planning process to one that revolves around multiple iterations through the development cycle.

The Agile movement was intended to address some of the challenges. In 2001, the "Agile Manifesto" was written by practitioners who proposed many of the Agile development methods. The manifesto states that Agile development should focus on four core values (www.agilemanifesto.org):

- Individuals and interactions over processes and tools.
- Working software over comprehensive documentation.
- Customer collaboration over contract negotiation.
- Responding to change over following a plan

Agile methods are designed to use a minimum of documentation in order to facilitate flexibility and responsiveness to changing conditions, implying that less planning and more flexibility are used in agile projects than in traditional project management. Agile methods have become more and more common in technology projects since their development (Lindvall et al., 2002) because they directly address the challenges so often confronted in dealing with dynamic projects in changing environments. Table 6.1 outlines some of the key diffferences between Traditional and Agile approaches.

In a study of software development projects, Boehm discussed similar sorts of challenges that development teams routinely face (Boehm, 1996, p. 74). A too-detailed requirements document can have the following problems: 1) specifications that do not describe a deliverable as well as the prototype, 2) early specification of requirements results in gold plating (adding more features than required) because there will be no further opportunities to add/change functionality, and 3) solutions focus on a specific point in time although the requirements or environment are likely to change (Table 6.1).

It is important to note that Agile does not abandon front-end planning as part of the project development methodology. Coram and Bohner, for example, point out that Agile methods do require upfront planning (Coram & Bohner, 2005). Significant communication and working with the customer is needed to provide project requirements for the first release. Indeed, the critical point is to recognize that, in many ways, more planning is performed in Agile environments, though the planning is spread across the entire development cycle, rather than occurring in an up-front, one-off manner.

However, a balance between traditional methods and Agile methods is usually appropriate. Certain factors, such as the size of the project, safety requirements, and known future requirements, call for upfront planning even in Agile projects, whereas turbulent, high-change environments call for less upfront planning and a greater use of Agile methods. Boehm suggests there is a "sweet spot," which is dependent on project characteristics where the effort expended in initial planning pays off in

TABLE 6.1

Main Differences between Traditional Development and Agile Development after Dybå and Dingsøyr (2008)

	Traditional Development	Agile Development
Fundamental assumption	Systems are fully specifiable, predictable, and are built through meticulous and extensive planning	High-quality adaptive software is developed by small teams using the principles of continuous design improvement and testing based on rapid feedback and Change
Management style	Command and control	Leadership and collaboration
Knowledge management	Explicit	Tacit
Communication	Formal	Informal
Development model	Life-cycle model	The evolutionary-delivery model
Desired organizational form/structure	Mechanistic (bureaucratic with high formalization), aimed at large organizations	Organic (flexible and participative encouraging cooperative social action), aimed at small and medium sized organizations
Quality control	Heavy planning and strict control. Late, heavy testing	Continuous control of requirements, design, and solutions. Continuous testing

project success (Boehm, 2002). Too much or too detailed planning can result in wasted effort and too much plan rework, whereas not enough initial planning can result in project failure. In analyzing 1,386 projects, Serrador and Turner found an "inverted U" relationship between planning and project success; in terms of the effort (time) taken to plan comprehensively (Serrador & Turner, 2015). That is, they found that too much effort and time spent planning can have just as negative impact on project success just as can as too little.

I wrote a paper with Jeff Pinto to investigate these questions (Serrador & Pinto, 2015). This chapter is based on that study.

Our study demonstrated that Agile and iterative methods had been widely adopted for the purpose of managing projects. We asked respondents to indicate what percentage of their projects included some elements of Agile methods, based on a detailed description of these activities. Nearly 6% of total projects were completely or nearly completely Agile. Further, more than 65% of the original 1,386 projects reported having some Agile or iterative component, determined by the percentage of Agile methods they indicated occurred in their project.

We found that the greater the Agile/iterative approach reported, the higher the reported project success. This finding was confirmed through a one-way analysis of variance (ANOVA) procedure. For the two assessments of project success: efficiency and stakeholder satisfaction, for project success the ANOVA demonstrated strongly statically significant differences ($p < 0.01$) in project success by degree of Agile methodologies employed in those projects. In the case of the Efficiency Factor, a marginally significant difference ($p < 0.10$) was also found. As a result, we can see there is clearly a relationship between Agile use and the success factors for stakeholders.

At this point, we should define a key statistical method, the p-value. The p-value is the probability of obtaining a test statistic at least as extreme as the one that was actually observed. In other words, it is the probability that observed result was a fluke. A p-value of 0.1 indicates a 10% chance the result was random and not due to a true relationship. As is standard for most studies in economics and management, I will use $p < 0.05$ as my cut-off for results in most cases. Our results in general were much better than that cut-off. Note that the p-value is not the only important value in our analysis. A result can have a very good p-value but with such a low R^2 value that the relationship is so weak as to not be important. We tested for both factors.

In addition, the correlation between the level of Agile development and success is small but statistically significant. What is shown is that Agile methodologies are correlated with higher reported success for each category: overall project success, efficiency, and stakeholder success (Table 6.2). It is also interesting to note that projects with a high Agile percentage report mean upfront planning amounts similar to traditional projects. As noted by Dybå and Dingsøyr, if substantial planning is done during execution then Agile projects appear to do more planning overall than traditional projects (Dybå & Dingsøyr, 2008).

To continue the analysis, a new index was created that combined the results of the question on the degree of Agile in the project with the measure of how much planning was done in the execution phase. Replanning during execution is a feature of agile methodologies (Dybå & Dingsøyr, 2008). The following measure was defined that used both

TABLE 6.2

Correlation Analysis between Methodology Type and Success Factors

Correlations N=1386			
	Overall Project Success Rating	Efficiency Factor	Stakeholder Success Factor
Methodology type	p = 0.000	p = 0.022	p = 0.000

*p <.005 is statistically significant.

TABLE 6.3

Standard Linear Regression Analysis of Combined Agile Measure vs. Project Success Rating

	N	R^2	p-level
Combined Agile measure	1002	.019	0.00001

items as a measure of the "Agileness" of the project. The first is based on respondent assessment of how much Agile process is used in the project; the second on how much replanning was completed during execution (Table 6.3).

One can see that the p levels are very good, though R^2 values are low at 0.019. However, all analyses suggest that the use of Agile methods is positively associated with improved success.

We should probably discuss what is meant by R^2 at this point. The coefficient of determination (R^2) provides a measure of how well future outcomes are likely to be predicted by a model. For example assume you are you are shopping and have no credit cards, just $100 cash. The R^2 of the relationship between the amount of cash you have and maximum you can spend shopping is therefore 1.00. It is directly related to the amount of cash you have with you; the more you have the more you can spend. However, if you also have a credit card with a $400 limit on you, then the R^2 between your cash and the maximum you can spend is 0.20 ($100/$500). But since we are referring to social sciences and economics, it is not that simple. If we look at how much you will actually spend, the relationship will also need to factor in how good the merchandise or prices are at the store, whether you need money for dinner later, and how long until your next paycheck. Maybe the R^2 between the cash in your pocket and how much you will actually spend is much smaller, maybe

0.10 or 0.05. Therefore the amount of cash in your pocket may only be a 10% predictor of how much money you will spend shopping. Maybe it is less. To summarize, R^2 indicates how much one variable affects another. In social sciences $R^2 > 0.6$ indicates two variables are probably measuring the same thing. R^2 above 0.05 is significant though R^2 less than 0.05 can also be significant in certain circumstances.

In this case, Agile is associated with successful projects but only with a 0.019 (Or about 2% impact). That seems quite small, but we decided to continue with the analysis.

We next conducted an MHRA analysis on the data to find the true impact of the planning effort on success. MHRA is a technique for separating the effect of other project characteristics from the effect being studied. The net result was a maximum R^2 of 0.152 which is high enough to be significant for project success impacts. In addition, we found that the use of Agile appears to have a greater impact on stakeholder success factors than straight efficiency. This is largely in keeping with the stated goal of Agile methodologies: to focus on products that meet stakeholder needs above all else (Table 6.4).

Next, we examined the efficacy of using Agile methods by industry. It is possible to determine some interesting patterns emerging (see Table 6.5). Four industries showed statistical significance when regressing methodology type versus overall success measure: high technology, health care, professional services, and the category reported by participants as "other." This finding confirms earlier work that has observed that Agile is more widespread in the high tech and IT fields and, in fact, Agile was originally designed for that type of environment (Dybå & Dingsøyr, 2008). On the other hand, industries where there is less reported use of Agile methodologies – such as construction, manufacturing, and retail – do not show a statistically significant relationship.

TABLE 6.4

Comparison for R^2 from MHRA Analysis between Combined Agile Measure and Success Factors

N=1002	Stakeholder Success Factor	Efficiency Factor
R^2	0.152	0.096
	p = 0.089	*p = 0.083*

TABLE 6.5

Comparison of Means and Regression Results for Agile Success by Industry

	Valid N	Regression p Value vs. Success
Construction	23	0.184
Financial services	73	0.747
Utilities	23	0.584
Government	34	0.336
Education	10	0.084
Other	53	0.0002*
High technology	57	0.035*
Telecommunications	35	0.570
Manufacturing	42	0.726
Health care	24	0.017*
Professional services	22	0.034*
Retail	16	0.722
All Groups	412	0.007*

Notes
* - indicates statistical significance.

DISCUSSION

For a number of years now, Agile has been touted as a methodology for project planning and execution that addresses many of the failings with the traditional, cascade planning process. Out of the frustration of multiple practitioners was born the *Agile Manifesto* and its call to dramatically reconsider the means by which successful projects are managed in chaotic settings. While the ideas that underpin the Agile philosophy are attractive and logical, to date, what has been lacking is empirical validation. Is an Agile-managed project more likely to succeed that one that relies on traditional approaches?

We explored the efficacy of the Agile method through a comprehensive and large-scale empirical analysis of projects being developed with varying levels of agile approaches and their subsequent likelihood of success. Our findings suggest that there is research support for the application of Agile methodology in managing projects. That is, we found that the level of Agile used in a project does have a statistically significant impact on all three dimensions of project success, as judged by efficiency, stakeholder satisfaction, and perception of overall project performance.

This chapter reports on one of the first empirical studies of the efficacy of Agile methods for project success (and employing the largest dataset). Although Agile has been used for project planning for a number of years now, to date, the majority of research examining its usefulness has been anecdotal, single-case studies, or based on small sample sizes in single-organization or single-industry settings. The likelihood of project success, as measured by multiple perspectives (efficiency, stakeholder satisfaction) was shown to improve through the use of Agile. It was also found that the quality of goals and vision of a project are important to the success of projects as expected, but particularly for Agile projects. Further, though it has been adopted in multiple industries and across national borders, our findings suggest that it has achieved best success to date within certain settings; notably, high technology, healthcare, and professional service. All of which are heavy users of software and IT. Overall, our study confirmed the trends seen in industry: Agile processes can help deliver more successful projects.

NOTE

1 This chapter is adapted from: Serrador, P., & Pinto, J. K. (2013). Does Agile work?—a quantitative analysis of agile project success. *International Journal of Project Management, 33*(5), 1040–1051.

REFERENCES

Boehm, B. (1996). Anchoring the software process. *IEEE Software, 13*(4), 73–82.

Boehm, B. (2002, January). Get ready for agile methods, with care. *Computer, 35*(1), 64–69.

Collyer, S., Warren, C., Hemsley, B., & Stevens, C. (2010). Aim, fire, aim—project planning styles in dynamic environments. *Project Management Journal, 41*(4), 108–121.

Coram, M., & Bohner, S. (2005). The impact of agile methods on software project management. *Proceedings of the 12th IEEE International Conference and Workshops on Engineering of Computer-Based Systems* (pp. 363–370). Washington, DC, USA: IEEE Computer Society.

Dybå, T., & Dingsøyr, T. (2008). Empirical studies of agile software development: a systematic review. *Information and Software Technology, 50*(9), 833–859.

Hällgren, M. A.-O. (2005). Deviations, ambiguity and uncertainty in a project-intensive organization. *Project Management Journal, 36*(3), 17–26.

Koontz, H. (1958). A preliminary statement of principles of planning and control. *The Journal of the Academy of Management, 1*, No, 1(1), 45–61.

Lindvall, M., Basili, V., Boehm, B., Costa, P., Dangle, K., Shull, F., ... Zelkowitz, M. (2002). Empirical findings in agile methods. In D. Wells, & L. Williams (Eds.), *Extreme programming and agile methods—XP/Agile Universe 2002, 2418*, 81–92.

Ritter, T. (2007). Public sector IT projects have only 30% success rate. *CIO for Department for Work and Pensions.* Retrieved from http://www.computerweekly. com/blogs/public-sector/2007/05/public-sector-it-projects-have.html

Serrador, P., & Pinto, J. K. (2013). Does Agile work?—a quantitative analysis of agile project success. *International Journal of Project Management, 33*(5), 1040–1051.

Serrador, P., & Turner, J. R. (2015b). What is enough planning? Results from a global quantitative study. *IEEE Transactions on Engineering Management, 62*(4), 462–474.

The Standish Group. (2011). *CHAOS Manifesto 2011.* Accessed June 2011, The Standish Group. Retrieved from http://standishgroup.com/newsroom/chaos_manifesto_ 2011.php

Turner, J. R., & Cochrane, R. A. (1993). Goals-and-methods matrix: coping with projects with ill defined goals and/or methods of achieving them. *International Journal of Project Management, 11*(2), 93–102.

7

Managing Complex Projects

Neil Turner, Kate Davis, and Chantal Cantarelli
Cranfield School of Management, Cranfield, United Kingdom

WHAT MAKES PROJECTS COMPLEX TO MANAGE?

Despite significant investment in project management processes, training, and professional Bodies of Knowledge, organizations large and small often still struggle to complete their projects as initially planned. Improving success rates represents a huge economic opportunity yet persists as an ongoing challenge. The tools, techniques, and standards developed by and for project managers are undoubtedly valuable, yet managers consider them as necessary but not sufficient to ensure their projects go smoothly. It is the complexity of their work that appears to hinder delivery.

Without a clear definition, complexity can be an unhelpful term as it can mean different things to different people. Much effort has been expended debating the difference between 'complex' and 'complicated', and there is no universally accepted definition. A useful view is that if you can understand a system in terms of how its parts interact, it is complicated. A commercial jet aircraft, for example, requires expertise to design and manufacture but it follows known laws. Many can be built that have repeatable performance. It is a linear system and can be taken apart, serviced, and re-assembled multiple times. Flying the aircraft with passengers aboard, however, can be a different matter. Here, bad weather can disrupt the pilots' schedule or planned route, and once we introduce passengers (not to mention a wide range of other stakeholders), we can see that it can be a rather unpredictable, non-linear system that cannot be fully controlled. This is the point at which the system becomes complex, rather than merely complicated.

DOI: 10.1201/9781003502654-7

The idea that all project issues encountered can readily be resolved by the application of appropriate management techniques is unrealistic. A more pragmatic view is that problems will inevitably occur, and that these must be confronted. Much academic work has looked at project complexity (e.g., Baccarini, 1996; Geraldi et al., 2011; Petro et al., 2019). Ackoff (1979) calls such practical situations 'messes', and this terminology seems to resonate with managers, who recognize this as their day-to-day reality. For practitioners, finding 'simple' solutions to their challenges is unlikely to be realistic (or they would have been implemented already!), but a different way of thinking about the problems to allow thoughtful responses to be crafted can be beneficial. An important point is that 'absolute' complexity in this sense (i.e., represented by a number or a point on a scale) is often unhelpful. It is important to consider who is doing the work. An experienced manager with many major projects under her or his belt will have seen many of the problems, or variants of them, before. In comparison, a new manager and team with no previous experience may well struggle even on a seemingly 'low-complexity' project because everything is new to them, and every problem is being encountered for the first time. When thinking about project complexity, it is vital to consider not only the 'what' of the work, but also 'who' is doing it.

RESEARCH FINDINGS

Asking the simple question 'What makes your project complex to manage?' we identified three distinct ways in which the work can quickly get messy (Maylor et al., 2013):

- *'Structural complexity'*: technical and coordination challenges (e.g., a construction project with numerous suppliers, a tight timescale, and a constrained budget). The difficulty increases with the scale and number of people involved, the number of interdependencies, and the variety and pace of the work being performed. Good project management training can often help address these difficulties.
- *'Socio-political complexity'*: difficulties due to people, power and political issues with the project (e.g., a development project with

multiple stakeholders who all have quite different views on what they want the work to achieve). Difficult stakeholder issues can be harder to quantify than structural factors, but this does not diminish the very real effect they have on project delivery. These may require more interpersonal skills rather than 'book knowledge'.

- *'Emergent complexity'*: these problems come from the unknown and the unexpected aspects of the work (e.g., a new product design where the market is unclear and early customer feedback is likely to cause changes to the requirements). Difficulties can be down to limited knowledge (especially in the early stages), and any changes that may arise during the project. By definition these can be hard to foresee but thinking of potential issues and areas you are unsure of can be helpful in navigating the work (see also PMI, 2014).

We have worked with thousands of managers over the years, and this terminology has been shown to be helpful in clarifying the types of difficulties particular projects face, allowing different categories of 'complex projects' to be distinguished. The next issue, though, is to find ways to alleviate those complexities.

It is useful to think about the kind of actions that would be most suitable in solving or reducing the particular problems encountered (Maylor and Turner, 2017). For structural complexities, a *planning and control* approach drawing on the management tools and techniques available (see Maylor and Turner, 2022) is generally advisable. There is a wealth of valuable material to support managers here. For socio-political complexities, *relationship development* and face-to-face discussions with key people can be powerful in diffusing tensions. This is an important (and sometimes under-emphasized) management responsibility. Finally, emergent complexity can be lessened by supporting *flexibility*, and encouraging pragmatic responses from the team as the project unfolds (Antonacopoulou et al., 2022). These responses are not meant to be exhaustive, but indicate how a manager may choose to use her or his time depending on the kind of complexity encountered.

Practical workshop data with managers shows that management practices are in fact varied and are not necessarily limited to their 'corresponding' complexity on the diagonal. Actual responses cover all nine options of complexity and response. This is indicated via the examples in Table 7.1.

TABLE 7.1

Relating Complexities and Responses (adapted from Maylor and Turner, 2017:1086)

Complexity Response	**Structural**	**Socio-political**	**Emergent**
Planning and control	Initiating, planning, monitoring (e.g., applying earned value systems).	Develop a communications plan/establish a project board of stakeholders.	Risk management and change control
Relationship development	Prioritize communications with stakeholders.	Engage in teambuilding activities.	Socialize changes.
	Conduct project outreach activities.	Invest in social capital.	Increase informal communications.
Flexibility	Embrace changes from process.	Manage expectations of change.	Agile PM approaches.
	Enable parallel development.	Joint planning with stakeholders.	Entrepreneurial PM practices.

SO, WHAT'S TO BE DONE?

What does this mean for managers? Table 7.1 indicates that there may not be a 'right' way to identify a complexity response, but it does mean that several options are available. Published studies indeed show that all the elements are in fact readily observed (e.g., Baxter and Turner, 2023; Boehme et al., 2021).

When analyzing what may be done about a complexity, there are three possible outcomes. First, it may be possible to *resolve* the issue using one of the three response options. This is unlikely to be straightforward, and will likely require spending scarce resources. It is important to note that there could in fact be a potential solution, but it may remain unimplemented. For example, if the root cause of the complexity is that the organization is doing too many projects and that resources are too thinly stretched to be effective (a very common problem), then a way forward is to prioritize the projects and stop the ones below the cut-off to allocate resources more appropriately. Easy in theory, but this is fiendishly hard to do in practice and few organizations can manage this successfully. Nevertheless, managers and teams we have worked with are often surprised by how many of these complexities can in fact be addressed. If a problem cannot be completely resolved, there may well be steps that can be taken to *reduce* the impact. Again, this may well involve additional expenditure, but it does indicate that these problems are not insurmountable. We have found through multiple workshops that, perhaps surprisingly, over 80% of problems raised can in fact be resolved or reduced. Many complexities are 'self-inflicted' within organizations, and through openly raising the issues, solutions can be identified and implemented to alleviate the problems. Finally, some complexities just cannot be fixed. We term them *'residual'* complexities and may include issues that are outside the organization's control, such as a particularly difficult stakeholder, or a business environment that they cannot influence. Even though this is not resolvable, sharing awareness of the issue can be useful if it creates a wider understanding of the challenge and allows stakeholders' expectations to be managed more effectively. A sudden announcement of a major unexpected slippage or huge cost overrun is always unwelcome, but keeping others appraised of issues as they unfold is often better than the shock of major bad news.

WHAT TO DO NEXT? PRACTICAL STEPS

Complex problems, by definition, do not lend themselves to simple solutions. We have found that discussing issues (often informally) and exploring potential ways forward is a powerful way to uncover sensible options. The three-by-three layout in Table 7.1 is a helpful tool for analyzing issues or concerns. What sort of complexity is it (structural, socio-political, emergent)? What response might potentially help resolve or reduce it (planning and control, relationship development, flexibility)? This can offer a structured and helpful discussion format in a group setting. What would make sense in this context? What realistically would be achievable? It is also worth considering the level at which solutions can be implemented – is it within the control if the manager and the team, or at the program/portfolio level?

It is also a beneficial reflective tool (Cecchi et al., 2022). Looking back at a project or phase often focuses on the structural (spend, duration, requirements met), but discussing the socio-political is valuable too

('How well did we treat our stakeholders?' 'Did the team develop their skills?'). For emergent complexities, rather than acknowledge how we were derailed unexpectedly, instead ask how we could have responded more effectively. Future (different) disruptions will certainly occur, and we need to consider how we can be more responsive and resilient next time. These are valuable conversations. It is important, however, to ensure that lessons are shared across the organization. Successful practices need to be promoted, and ineffective responses identified so that they are not repeated. This is another notoriously difficult problem, but a combination of formal and informal knowledge-sharing techniques can be used to boost project performance over time – a hugely valuable outcome.

In the end, though, complexity management requires the skills and expertise of the manager and the team. There is no one-size-fits-all panacea, and solutions cannot be outsourced to a convenient tool or process. Situated human judgment remains key, and this is what separates the great projects from the merely good.

REFERENCES

Ackoff, R. 1979. 'The future of operational research is past', *Journal of the Operational Research Society*, 30(2): 93–104.

Antonacopoulou, E., Turner, N., Al-Tabbaa, O., Michaelides, R. and Schuster, A. 2022. 'Advancing Temporal Organizing: The Case for a Practising School in Project-Based Organizing', *Academy of Management Conference*, Seattle, Washington, 5–8 August.

Baccarini, D. 1996. 'The concept of project complexity – a review', *International Journal of Project Management*, 14(4): 201–204.

Baxter, D. and Turner, N. 2023. 'Why Scrum works in new product development: The role of social capital in managing complexity', *Production Planning & Control*, 24(13): 1248–1260.

Boehme, T., Aitken, J., Turner, N. and Handfield, R. 2021. 'Covid-19 response of an additive manufacturing cluster in Australia', *Supply Chain Management: An International Journal*, 26(6): 767–784.

Cecchi, M., Grant, S., Seiler, M., Turner, N., Adams, R. and Goffin, K. 2022. 'How COVID-19 impacted tacit knowledge and social interaction of global NPD project teams', *Research-Technology Management*, 65(2): 41–52.

Geraldi, J. G., Maylor, H. and Williams, T. 2011. 'Now, let's make it really complex (complicated): A systematic review of the complexities of projects', *International Journal of Operations & Production Management*, 31(9): 966–990.

Maylor, H. and Turner, N. 2017. 'Understand, reduce, respond: Project complexity management theory and practice', *International Journal of Operations and Production Management*, 37(8): 1076–1093.

Maylor, H. and Turner, N. 2022. *Project Management*, 5th edition. Harlow, Pearson.

Maylor, H., Turner, N. and Murray-Webster, R. 2013. 'How hard can it be? Actively managing complexity in technology projects', *Research-Technology Management*, 56(4): 45–51.

Petro, Y., Ojiako, U., Williams, T. and Marshall, A. 2019. 'Organizational ambidexterity: A critical review and development of a project-focused definition', *Journal of Management in Engineering*, 35(3): 1–20.

PMI. 2014. *Navigating Complexity: A Practice Guide*. Newtown Square, PA.

8

Managing the Benefits of Collaborative Projects[1]

David O'Sullivan[1] and Gabriela Fernandes[2]
[1]University of Galway, Galway, Ireland
[2]University of Coimbra, Faculty of Sciences and Technology, Coimbra, Portugal

INTRODUCTION

We know projects as temporary endeavours with the aim of achieving specific goals within the constraints of scope, time, and resources. Project management is about planning, executing, and controlling projects in a way that meets these goals with high quality, efficiency, and effectiveness. Goal seeking is a key trait of every open system. Other traits include holism, hierarchy, transformation, and entropy. Open systems also have equifinality and feedback (Boulding, 1956). Equifinality means that there are multiple ways of achieving the same goals. It means that projects exhibit a degree of flexibility and adaptability, allowing initial causal factors to reach goals through different trajectories. Additionally, project goals can be multifaceted and can extend well beyond the triple constraints of time, resources, and scope and into constructs such as objectives, needs, benefits, values, impacts, and their synonyms.

Collaborative projects involving two or more distinct organisations complicates our need to understand 'goal seeking' further. Collaborative projects introduce two or more cultures of partner organisations with their own requirements and understanding of goals. This typology of collaborative projects can also be holistically and hierarchically complex, with multiple interdependent goal-seeking subsystems. Collaborative projects are typically of long duration and more likely to require actions that combat the effects of entropy. Collaboration also increases the need for good feedback and communication that enhance trust building between partners. Collaborative projects also have different sets of end users or customers. Collaborating organisations may find it difficult to directly access the goal or 'voice of the customer' of their partner organisations.

PROJECT OBJECTIVES VS. PROJECT BENEFITS

Project objectives are a specific articulation of a type of goal and normally refer to specific outcomes necessary to achieve at the successful completion of a project during its lifecycle. Collaborative project objectives are multi-faceted. Identifying and understanding project objectives is a crucial first step in planning and managing the project effectively. In this chapter project objectives refer to specific, tactical, or short-term outcomes such as meeting time, cost, and scope objectives during the lifetime of the project. Objectives are specific, measurable, achievable, relevant, and time bound. Project objectives provide a clear understanding of what the project team is expected to accomplish and serve as a basis for evaluating short-term project success. Project objectives are specific and measurable so that progress can be tracked, and success can be evaluated during the lifetime of the project. They are achievable and realistic, given the available resources and constraints. Finally, project objectives are time-bound, with clear deadlines and milestones, so that progress can be monitored, and adjustments can be made if necessary.

Project benefits are different to project objectives. Project benefits refer to general, high-level objectives that meet or contribute towards stakeholder expectations and strategic gain in the long-term and beyond the life of the project. Project benefits are strategic and linked to the overall business strategies of the stakeholders beyond the immediate project objectives. They represent the value that the project will provide to the organisation and its end-users in the long term. Project benefits are typically qualitative statements but can be accompanied with key quantitative performance indicators.

As a rule of thumb, if a project benefit does not look and feel like it will remain relevant beyond the lifecycle of the project, then they should be regarded as project objectives and vice versa. Project benefits are identified and quantified during the planning phase of the project and monitored throughout the project lifecycle. They should be aligned with the overall strategic goals of each of the stakeholders. In addition, project benefits should be regularly communicated to all relevant parties to ensure buy-in and support for the project. By focusing on project benefits, organisations can ensure that their project delivers long-term value and contribute to the overall success of the organisation. In collaboratives projects, project benefits also reduce ambiguity among stakeholders and provide common language for partners from different cultures.

BENEFITS MANAGEMENT

Inter-organisational or collaboration projects, such as research and development (R&D) consortia between universities and industry, pose unique challenges that include high uncertainty, diverse organisational cultures, distinct benefits requirements for each partner, and the involvement of multiple stakeholders pursuing different and sometimes competing benefits. To address these additional challenges, a framework specific to inter-organisational collaboration projects was proposed by Fernandes and O'Sullivan (2021). Using the existing benefits management frameworks, as a starting point and adopting an open systems approach, they conceptualised the framework presented in Figure 8.1 as a roadmap for how to manage benefits in large collaborative projects.

This benefit management framework comprises four key activities (1) Identify Expected Benefits, (2) Plan Benefits Realisation, (3) Pursue Benefits Realisation and (4) Transfer and Sustain Benefits. These are the four parent activities of a larger set of child activities in a so-called SADT model (Ross and Schoman, 1977) – a technique that models many of the traits of open systems and that can be applied to any open system including projects. We will now review each of these four key activities in turn.

IDENTIFY EXPECTED BENEFITS

The first activity on the benefits management framework is to Identify Expected Benefits and as the name suggests, is about gathering information and knowledge about the kinds of benefits that can lead,

Project
Planning

Project
Initiation

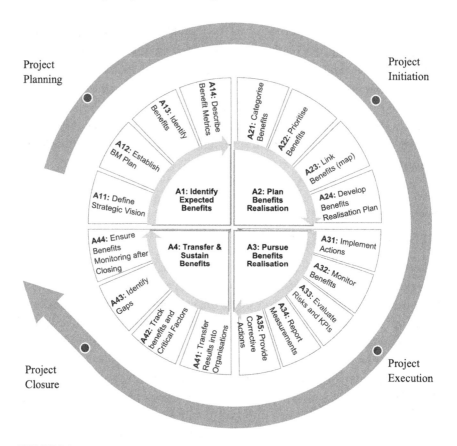

Project
Closure

Project
Execution

FIGURE 8.1
Benefits management framework.

inform, inspire, and motivate the project team. A first step in this activity is to define a benefits vision for the outcomes from the project. Vision is a concise statement that outlines the projects long-term goals in a single statement. Although aspirational, the statement must also be credible and realistic. It serves as a guide for decision-making, setting objectives, and communicating the purpose of the project various stakeholders. Key attributes of a project vision include relevance i.e., ensuring that the statement reflects the core purpose and goals of the project. It should also be realistic and achievable within a reasonable timeframe. Vision needs to inspire and challenge the project team without being unattainable. It should also be aligned with capabilities not only of each stakeholder but also the project team and ideally build on their unique strengths. Vision should be consistent with the values of the stakeholders and where possible reflect the cultures of each consortia partner. Finally, vision should consider the perspectives and expectations of key stakeholders. An example of a project vision statement for a major university–industry collaboration project is presented below for illustration:

> *Our vision is to create a transformative university–industry collaboration that drives innovation, fosters knowledge exchange, and generates positive societal impact. Through our partnership, we aim to be a catalyst for breakthrough advancements, empowering students, researchers, and industry professionals to co-create cutting-edge solutions.*

The second step in Identify Expected Benefits is to establish the benefits management plan or more accurately the workflow for a proposed plan, since the plan will contain multiple stages, activities, and tools and techniques. This plan should clearly be aligned with the benefits management framework but also highlight the key tools and techniques that will be deployed throughout the project lifecycle and beyond. One such workflow (or sample plan) is presented in Figure 8.2. This sample workflow can be customised for each project according to the language, customs, and practices of the stakeholders. The sample plan identifies three main tools in the project planning phase – the benefits Vision

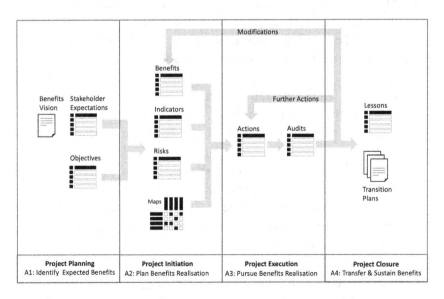

FIGURE 8.2
Benefits management plan workflow.

Statement, a Stakeholders Expectations list, and the project Objectives list. In another project these tools might be labelled mission, goals, or requirements respectively and may include additional tools such as a Benchmarks list and a strengths, weaknesses, threats, and opportunities (SWOT) list.

The third step of this first phase of the benefits management framework is Identify Benefits and as the name implies is the process of determining initial stakeholder expectations and initial project objectives. The activity should be supported by a benefit breakdown structure to act as benchmarks in the development of agreed benefits among different stakeholders (Fernandes et al., 2020a). This is followed by the final step to Describe Benefits Metrics or more particularly to identify Indicators of performance for both the project objectives and the proposed long-term benefits.

PLAN BENEFITS REALISATION

The second activity in the Benefits Management Framework is Plan Benefits Realisation. This second activity has four child activities: Categorise Benefits, Prioritise Benefits, Link Benefits, and Develop Benefits Realisation Plan. Building on the hypothetical vision statement presented earlier, and for illustration only, the following are sample benefits for a large university-industry collaboration:

- Exchange knowledge and expertise.
- Leverage the resources, facilities, and specialised skills.
- Transfer cutting-edge technologies and innovations between partners.
- Offer collaborating partners access to specialised training programs.
- Attract top-tier students, researchers.
- Enhance industry competitiveness by leveraging academic research.
- Drive regional and national economic growth through entrepreneurship.

Categorise Benefits is the process of categorising groups of benefits, by using a benefits breakdown structure. The Balanced Scorecard (Kaplan and Norton, 2005), for example, would categorise benefits under four headings or perspectives i.e., Financial, Customer, Internal Processes, and Learning & Growth. The Financial group would focus on financial benefits such as revenue growth, profitability, return on investment, and cost reduction. The Customer heading groups benefits related to customer satisfaction, retention, acquisition, and market share. The Internal Processes heading focuses on benefits for the internal processes and activities that are critical for achieving customer satisfaction and financial success. Finally, the Learning & Growth grouping looks at the organisation's ability to learn, develop, and adapt. It includes benefits related to employee training, skills development, employee satisfaction, and organisational culture. It's important to emphasise that each project consortia would need to develop their own categorisations.

Prioritise Benefits can be as simple as scoring the importance of each individual benefit. A simple method such as allocating a number from one to five for 'relevance' and the same for 'importance' yields a list of prioritised benefits. The assumption is that some benefits are more important than others and so would receive more of the attention and effort.

Link Benefits involves mapping benefits against, for example, Stakeholder Expectations. This linking, using, for example, a simple matrix would illustrate which benefits were not linked to Expectations or which expectations had no benefits. This exercise would prompt the need for defining new or revised benefits. A simple Benefits Map is illustrated in Figure 8.3 for a matrix between Benefits and Stakeholder Expectations

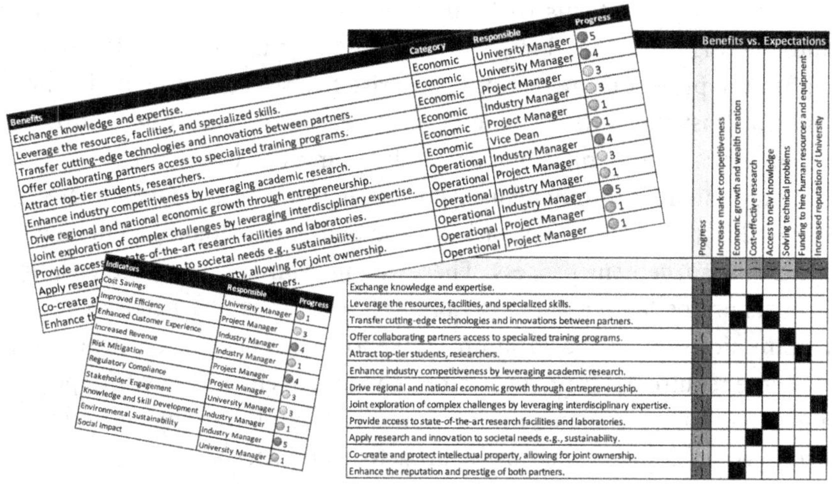

FIGURE 8.3
Sample of benefits, indicators, and mapping.

and includes a simple graphical way of visualising progress over time using smiley faces.

The final child activity is Develop Benefits Realisation Plan and brings together all the elements discussed so far i.e., Expectations, Objectives, Benefits and Maps. Other elements could also be included, for example, key performance Indicators. Indicators is a list of key performance indicators (KPIs) related to measuring the benefits portfolio, and in particular progress of these Indicators over the lifetime of the project and beyond. Reiterating a point made earlier, each project consortia would need to develop their own agreed workflow and tools that make sense within their own individual cultures and ways of working.

PURSUE BENEFITS REALISATION

The third activity in the Benefits Management Framework is Pursue Benefits Realisation. This third activity has five child activities: Implement Actions, Monitor Benefits, Evaluate Risks and KPIs, Report Measurements, and Provide Corrective Actions. This part of the Benefits Management framework is action oriented, i.e., what are the large and small actions necessary to steer the project towards its vision of long-term benefits. Actions are continuous improvement efforts, with their own due dates, leaders, and resources. They are mini projects within a project whose sole focus is the development and realisation of benefits and making sure that long-term benefits remain a focus for change within the project team. Actions can arise from several potential sources and not least the benefits defined in the earlier phase. Other sources of actions include risk management and recommendations from Audits.

Actions arising from Benefits: Actions refer to the steps taken to address issues, deviations, problems, or ideas and endeavours identified during the benefits management process. These actions can ensure that the desired benefits are better achieved. Actions can arise from the need for greater communication or clarification, stakeholder concerns or aspirations, resource requirements or constraints, changes of enhancements to policies and procedures, revising project objectives, and training and support needs. Actions should be timely, proactive, and specific.

Actions arising from Risk Management: A risk management plan is a systematic approach to identifying, assessing, and managing potential risks and uncertainties that could impact benefits realisation. It contains several key elements including identified risks, the likelihood and potential impact of each risk, and actions that can be carried out to avoid the risk from occurring. The project team can then focus only on actions that mitigate against the most likely and most severe risks occurring. Risk management includes activities include regular monitoring, communication, and reporting. A risk management plan and the Actions that arise, proactively identifies, and addresses potential risks, enhance decision-making, and increase the likelihood of achieving projected benefits.

Actions arising from Audits: Audits or reviews can be a key part of the benefits management process. Self-assessment is a process of the project team and stakeholders taking time to review various strengths, weaknesses, and potential enhancements of a project including benefits management. Independent assessment, on the other hand, is typically carried by individuals unrelated to the collaborative project team. They may be members of the partner organisations and include external domain experts and facilitators. The purpose is to review or audit various aspects of the project and then suggest enhancements that can be converted into Actions for implementation.

TRANSFER AND SUSTAIN BENEFITS

The final activity in the Benefits Management Framework is Transfer and Sustain Benefits. This final activity has five child activities: Transfer Results into Organisations, Track benefits and Critical Factors, Identify Gaps, and Ensure Benefits Monitoring after Closure. This part of the Benefits Management framework is mainly focused on communications and learning. Two key tools are illustrated in Figure 8.3 for this phase, Lessons Learned and Transition Plans.

Lessons Learned or post-project reviews is a process of documenting insights on the project's successes, challenges, failures, and valuable lessons that can be applied to future projects. The process usually involves workshops among key participants who can brainstorm and openly share positive and negative insights. A dedicated session or series of sessions may be necessary to review a large project. Participants will include key stakeholders, team members, and relevant parties who were directly or indirectly involved in the project. It may also involve an independent domain expert and facilitators or observers. Lessons Learned sessions are carried out in an open and non-judgmental atmosphere.

Transition Plans can be another important tool in facilitating ongoing benefits realisation. A project transition plan outlines the steps and activities necessary to smoothly transfer a project into the mainstream operational processes. In essence, transition plans are mini projects within each partner organisation with their own objectives, project team, schedules, and deliverables. Transition objectives establish the desired outcomes and success criteria for the transition. Knowledge transfer involves determining the knowledge, information, and documentation that need to be transferred. This includes transferring lessons that should be shared with the receiving party. A communication plan can ensure effective and timely communication during the transition, the frequency and methods of communication, and the key lessons to be delivered. Other key elements of a Transition Plan include testing and validation criteria and support and training.

CONCLUSIONS

Project objectives are specific outcomes necessary for project success. They cover scope, resources, time, cost, quality, risk, and stakeholder expectations. Collaborative projects require coordination, collaboration, and planning to deliver objectives and align with stakeholders' strategic goals. Project benefits differ from objectives and refer to high-level, long-term objectives contributing to stakeholder expectations and strategic gain. Benefits are both qualitative and quantitative, representing the value that a project provides in the long term. Benefits should align with stakeholders' strategic goals, be communicated regularly, and reduce ambiguity among stakeholders.

The Benefits Management Framework for collaborative projects includes key activities like identifying expected benefits, planning benefits realisation, pursuing benefits, and transferring and sustaining benefits. Identifying Expected Benefits involves defining a strategic vision, developing a benefits management plan, gathering stakeholder expectations and project objectives, and describing benefit metrics.

Planning Benefits Realisation includes categorising and prioritising benefits, linking them to project objectives, and developing a benefits realisation plan. Techniques like workshops and focus groups aid in engaging stakeholders and gathering their expectations. Pursuing Benefits Realisation is action oriented and gets its actions from the benefits list but also risk management and internal and external audits. Transferring and Sustaining Benefits is principally about learning lessons from the project in terms of benefits management and considering transitions plans to transfer the responsibility for benefits management from the project team to organisations' key stakeholders.

In summary, project objectives focus on specific outcomes, while project benefits encompass long-term objectives. Benefits management frameworks and workflows provide guidance for managing benefits, and specific frameworks address challenges in inter-organisational collaboration projects. In this chapter we have advocated for a strategic shift from the triple constraints of time, cost, and quality to more strategic measures. We have emphasised the need for a tailored benefits management process for each project context, with the framework and sample workflow presented above serving as a starting point. Finally, it is worth mentioning that a Project Management Office or similar structure plays a central role in embedding the benefits management by supporting stakeholders in each step of the benefits management process (Fernandes et al., 2020b).

NOTE

1 This research is sponsored by national funds through FCT – Fundação para a Ciência e a Tecnologia, under the project UIDB/00285/2020 and LA/P/0112/2020.

REFERENCES

Boulding, K.E. (1956). General systems theory—the skeleton of science. *Management Science*, 2(3), 197–208.
Fernandes, G., & O'Sullivan, D. (2021). Benefits management in university–industry collaboration programs. *International Journal of Project Management*, 39(1), 71–84.
Fernandes, G., Araújo, M., Andrade, R., Tereso, A. Pinto, E.B., & Machado, R.J. (2020a). Critical factors for benefits realisation in collaborative university–industry R&D

programmes. *International Journal of Project Organisation and Management*, 12(1), 1–30.

Fernandes, G., Pinto, E.B, Araújo, M., & Machado, R.J (2020b). The roles of a programme and project management office to support collaborative university–industry R&D. *Total Quality Management & Business Excellence*, 31(5–6), 583–608.

Kaplan, R.S. and Norton, D.P. (2005). *The balanced scorecard: measures that drive performance* (Vol. 70, pp. 71–79). Harvard Business Review.

Ross, D.T. and Schoman, K.E., Jr. (1977). Structured analysis for requirements definition. *IEEE Transactions on Software Engineering*, SE-3(1), 6–15.

9

Agile, Traditional, and Hybrid Approaches to Project Success [1]: Is Hybrid a Poor Second Choice?

Andrew Gemino and Blaize Horner Reich
Beedie School of Business, Simon Fraser University, Vancouver, British Columbia, Canada

INTRODUCTION

Researchers and practitioners continue to explore important factors that impact project success. One potential factor, the use of a project management approach or methodology, remains relatively unexplored. This factor becomes critical as new project approaches, such as Agile, emerge and challenge existing theories and practices, especially in the field of software and IT-enabled business projects. An additional complexity is that project approaches are being combined into novel hybrid forms.

Our research (Gemino, Reich, and Serrador, 2021) was one of the very few that empirically tested the impact of project management approach using multiple measures of project performance.

To start, we established common definitions to be used to distinguish between a project management approach, methodology, and practice. A project management approach is an overall set of principles and guidelines which define the way a specific project is managed. Common approaches are traditional (aka waterfall, planned), and agile. A project management methodology is a system of techniques, procedures, and rules used by those that work in the discipline. Examples of project management methodologies might include Prince 2, Scrum, and Kanban. Within methodologies are the practices used to manage an aspect of a project. Examples of project management practices would

DOI: 10.1201/9781003502654-9

include daily stand-ups, Kanban boards, work breakdown structures and risk registers.

Traditional, Agile, and Hybrid Approaches

For decades, projects have been managed using collections of practices and methodologies that we refer to as the Traditional approach. The Traditional approach is defined by linear and predictable project planning practices designed to achieve a well understood, achievable set of objectives. As researchers have noted, *"The ultimate goal of the Traditional project management approach is optimization and efficiency in following [the] initial detailed project plan, ... to finalize [the] project within planned time, budget, and scope"* (Spundak, 2014, p. 941).

For the past two decades, Agile approaches have emerged into the mainstream. There are many Agile methodologies and, within them, Agile practices, but they are unified by a common conceptual foundation which can be called an Agile approach. Key among these foundations is the ability to adapt to changes and divide the work into distinct iterations throughout the project. Another critical element in an Agile approach is the distribution of responsibility to team members and the inclusion of project stakeholders (particularly clients and sponsors) in both formal and informal communication around the project.

A third approach, which combines Traditional and Agile, is emerging. At its most general level, a "hybrid" project management approach combines methodologies and practices from more than one project management approach. In our research, a hybrid approach is combining agile and traditional practices. An early research study (Serrador and Pinto, 2015) found that 62% of projects were neither fully agile nor fully traditional, which led us to examine the success of hybrid versus the more established project approaches.

Why organizations and project managers are using Hybrid approaches has not been rigorously studied to date. There seem to be two views, depending on whether the author starts from an Agile or a Traditional perspective. Studies starting from the Agile perspective call the phenomenon Hybrid Agile and discuss the difficulties organizations have in adopting a pure Agile approach. Some of the reasons that a pure agile approach may not be possible are pre-set budget and time constraints, regulatory requirements, and governance concerns.

Organizations which start from a Traditional approach may add Agile practices to solve a known issue in their project context. For example, the ABC CruiseLine (Batra et al., 2013) added Agile practices when the sponsors realized that the project was likely to encounter significant changes in user requirements. Others may adopt certain Agile techniques – for example, daily stand-up meetings or shorter iterations – to improve their outcomes.

RESEARCH SAMPLE

Each of the 477 survey responses in our sample represented a single completed project and the respondent reported on project characteristics, project success, project methodologies and practices, and other factors that may have impacted the project. Respondents were experienced project managers. Characteristics of the sample are shown in Table 9.1.

Measures

The three sections in this chapter describe how the research measured project success, project management approach, and the independent variables. Where possible, measures utilized items developed in previous research to provide external validity.

TABLE 9.1

Characteristics of the Survey Sample

Characteristic	Measure
Number of projects in the sample (N)	477
Years of PM experience of respondents	13.7 years (mean), 12 years (median)
Project budget	$3.46 M US$ (mean), $450 K (median)
Project duration	13.7 months (mean), 12 months (median)
Project team size (full time equivalents)	22 people (mean), 10 people (median)
Gender of respondents	80% male, 20% female
Project type	65% software, 35% other
Location of project	61% N. America, 39% other

TABLE 9.2

Measures of Project Performance

Measure of Project Performance	Survey Questions
Budget & Time Success	How successful was the project in meeting project budget goals?
	How successful was the project in meeting project time goals?
Scope & Quality Success	How successful was the project in meeting scope and requirements goals?
	How do you rate the quality of the project deliverables?
Stakeholder Success	How did the project sponsors and stakeholders rate the success of the project?
	How do you rate the client's satisfaction with the project's results?
	How do you rate the project team's satisfaction with the project?

Measuring Project Performance

We asked multiple questions that related to various aspects of performance and three distinct measures of success emerged from analysing the data – budget/time, scope/quality, and stakeholder success. Table 9.2 below shows the questions to measure each aspect of performance.

Measuring the Project Management Approach

Rather than asking respondents directly what approach they used in their project, we had a survey question asking them to identify the percentage of Agile practices used in their project (choices were 0–19%, 20–39%, 40–59%, 60–79%, and 80–100%).

If respondents selected between 80% and 100% Agile practices, the project was categorized as an Agile Project. If they selected between 0% and 19% Agile practices, the project was categorized as a Traditional project. If the percentage of Agile practices fell between 20% and 79%, the projects were categorized Hybrid. As the definition of Hybrid is "combining practices from Agile and Traditional approaches", we considered 20% as a reasonable cut-off for substantial use of the other approach.

TABLE 9.3

Independent Variables – Survey Questions, Measures, and Representative Sources

Item/Variable	Survey Question Measured from 1= Very Low to 5 = Very High
Project complexity	Rate the complexity of the project
Goal clarity	Rate the clarity of the vision statement or project goal definition for the project
Team experience level	What was the experience level of the project team?
Team attitude to change	How well did the team and the project respond to external changes?
Top management support	How supportive was senior management of this project and goals?
Degree of stakeholder engagement	How engaged were the key stakeholders for the project?

When testing for the impact of a specific variable (e.g., project management approach), it is important to identify the unique contribution of the factor, over and above other factors that are known to be influential. To accomplish this, we asked questions about six variables: project complexity, goal clarity, project team experience, project team attitude to change, top management support, and stakeholder involvement. Survey questions are shown in Table 9.3.

RESULTS

Of the 477 projects in the sample, 52% were classified as using a hybrid approach, 33% used a traditional approach, and 15% used a pure agile approach. This distribution is similar to findings from prior research.

We developed three regressions that used project approach (Traditional, Agile, Hybrid) and six independent variables (Complexity, Clarity, Team Experience, Change Attitude, Management Support, and Stakeholder Engagement) to predict three project success variables: (1) Budget & Time Success, (2) Scope & Quality Success, and (3) Stakeholder Success. All the regressions were significant with relatively strong adjusted R^2. Adjusted R^2 provides an estimate of the percentage of variation explained by each regression model (see 1, 2, and 3 in Table 9.4).

TABLE 9.4

Overall Regression Results

Overall Regression – All Independent Variables and Project Approach Variables Included			
Project Performance Variable	Adjusted R^2 % of Variance Explained	F Stat Test for Overall Significance	Significance Level (.000 highest)
1 Budget & Time Success	0.214	16.9	0.000
2 Scope & Quality Success	0.268	22.4	0.000
3 Stakeholder Success	0.405	40.9	0.000

Impact of Project Management Approach on Project Success

The primary question of interest in this study was to consider whether the project approach taken to manage a project had a significant effect on project success.

The data show that projects using Agile, Hybrid or Traditional approaches exhibit similar levels of performance on the "iron triangle goals" of (1) Budget & Time and (2) Scope & Quality. This is an interesting finding as it suggests that there is no reason to predict inferior performance on Budget & Time, or Scope & Quality goals from projects which are either mostly or partially using Agile practices. This result may address practitioner concerns that Agile practices are more expensive, require more time or lead to additional increases in scope. These results, provided in Table 9.5, show no significant negative effect of using Agile methodology and practices.

Another finding is that Agile and Hybrid approaches significantly and strongly outperformed Traditional on the Stakeholder Success measure. This is shown in the bottom right corner of Table 9.5. Both Agile and Hybrid approaches increased Stakeholder Success by very significant margins (i.e., 0.049 and 0.044, respectively), in addition to the contribution of the other variables tested.

It is not unexpected that an Agile approach would outperform a Traditional approach to deliver Stakeholder Success. In this study, Stakeholder Success is measured as a combination of sponsor, client, and team satisfaction with the project. A fundamental tenet of the Agile approach is frequent communication between these three stakeholders and the positive impact that successful communication can have on a

TABLE 9.5

Impact of Agile and Hybrid on Success Factors

Dependent Variable (Performance Measures)	Intercept Size of Effect Significance	Agile Size of Effect Significance	Hybrid Size of Effect Significance
1 Budget & Time	0.384 significant	−0.002 NOT significant	0.013 NOT significant
2 Scope & Quality	0.305 significant	−0.004 NOT significant	0.030 NOT significant
3 Stakeholder Success	0.150 significant	0.049 significant	0.044 significant

project (Tessem, 2018). At the end of an Agile project, there should be no important gaps between what was expected and what was delivered, and stakeholders should have reconciled themselves to the outcome. In a Traditional approach which features less communication throughout, surprises can surface later in the project, leaving sponsors and teams dissatisfied. Therefore, we would expect Agile to outperform Traditional on this metric.

The interesting result was that the Hybrid approach was very strongly positively impactful on stakeholder success. Mixing Traditional and Agile practices intentionally and mindfully seems to work well.

There is little theory yet about how project managers combine practices and therefore how well Hybrid projects should perform. If Hybrid projects are all fundamentally Agile projects with a few traditional practices like WBS or risk management added, then they might be expected to perform well on Stakeholder Success. However, if they are fundamentally Traditional projects with team stand-ups or a Kanban board added, it isn't easily predictable how they would perform.

We believe that project managers using a Hybrid approach are incorporating at least one of the fundamental Agile principles, specifically that businesspeople and developers work together throughout the project. This principle is often stated from a software development project perspective but can be implemented on any type of project. This practice would positively impact Stakeholder Success over Traditional approaches, which do not require or even encourage frequent interaction between the team, sponsors, and clients.

Our over-arching conclusion from this study is that a thoughtfully developed and well executed approach to a project, whether it is Agile,

Traditional, or Hybrid, can produce predictable triangle results. To produce superior results on Stakeholder Success requires a higher level of communication between the team, sponsor, and client, one that is delivered using Agile principles.

Future of Hybrid Project Management

This research demonstrates that a Hybrid project management approach can provide the same Budget & Time and Scope & Quality outcomes commensurate with Traditional approaches, while at the same time attaining the same level of Stakeholder Success that Agile approaches deliver.

Quantitative and qualitative research and standards bodies are predicting that hybrid approaches are prevalent and a natural choice for project managers. The conditions that lead to the adoption of Hybrid approaches are unlikely to disappear; we expect them to be a significant if not the dominant way that projects are conducted for the foreseeable future.

IMPLICATIONS FOR PRACTICE

We examined the impact of three different project management approaches – Agile, Traditional, and Hybrid. Results from the survey indicated that the Hybrid approach was more widespread than expected, representing approximately 50% of the sample. Not surprisingly, Agile approaches outperformed Traditional approaches in delivering Stakeholder Success. More surprising were the findings that Hybrid approaches also outperformed in delivering Stakeholder Success and that there were no significant differences between any of the approaches with respect to achieving Budget & Time or Scope & Quality success. There are three conclusions that sponsors, and senior stakeholders of projects could take away from this study.

When considering organizational practices, sponsors and senior managers could seek to free projects from the "tyranny" of methodologies and instead challenge project teams to use practices that add value or match their environment regardless of their methodological origins. As noted in West et al. (2011, p. 14) *"The result will be a more robust, flexible* process that evolves in response to the situation rather than a well-documented, inflexible process that assumes that all problems are the same."

A second contribution comes from secondary findings of this study. Reducing complexity, creating clear scope requirements, and maintaining a high level of stakeholder engagement had positive effects on all measures of project success. This strong result further supports the practice of combining Traditional approaches (i.e., establishing clear scope requirements) with the essential element of Agile approaches (i.e., maintaining high levels of stakeholder engagement, delivering results iteratively). This again points to the logic of a Hybrid approach.

In our opinion, Hybrid approaches are not a stage that organizations go through as they move from one project management approach to another. In bringing elements from different approaches together to create more successful outcomes, a Hybrid approach represents a maturing of the project management discipline.

NOTE

1 Portions of this chapter were adapted from: Gemino, A., Reich, B.H. & Serrador, P. (2021) Agile, Traditional, and Hybrid Approaches to Project Success: Is Hybrid a Poor Second Choice?, Project Management Journal, 52(2), 161–175.

REFERENCES

Batra, D., Xia, W., VanderMeer, D., & Dutta, K. (2013). Balancing Agile and Structured Development approaches to successfully manage large distributed software projects: A case study from the cruise line industry. *Communications of the Association of Information Systems*, (27).

Gemino, A., Reich, B. H., & Serrador, P. (2021). Agile, Traditional, and Hybrid approaches to project success: Is Hybrid a poor second choice? *Project Management Journal*, 52(2), 161–175.

Serrador, P., & Pinto, J. K. (2015). Does Agile work? A quantitative analysis of Agile project success. *International Journal of Project Management*, 33(5), 1040–1051.

Špundak, M. (2014). Mixed Agile/Traditional project management methodology–reality or illusion? *Procedia-Social and Behavioral Sciences*, 119, 939–948.

Tessem, B. (2018). What causes positive customer satisfaction in an ineffectual software development project? A mechanism from a process tracing case study. *Project portfolio risk categorisation–factor analysis results*.

West, D., Gilpin, M., Grant, T., & Anderson, A. (2011). Water-scrum-fall is the reality of agile for most organizations today. *Forrester Research*, 26.

10

Deadline. Dead-Line. Breathing Life into Projects

Jonas Söderlund
Linköping University, Linköping, Sweden

A WORD ON LIFE

Think about the word "deadline" – "Dead" "Line" – a common word used in a number of project settings. In projects, there is always some kind of deadline – a time limit, a particular date when things should be done. Of course, deadlines exist in all kinds of management settings – but in projects, they are often the most important part of management, the most important part of the existence for the entire project. Deadlines are a key characteristic for project managing and, in that respect, it is essential for project managers to reflect on the role and effects of deadlines. Are we using deadlines in the right way? How can we use deadlines even better – to trigger collaboration and creativity? My experience is that deadlines are too important not to be taken seriously and project managers tend to use deadlines casually and unreflectively.

I have always been interested by the phenomenon of the deadline. In fact, this was the very start of my own research into the management of projects. Deadlines became my way into understanding the rationale of organizing by projects and addressing what I believe are the very core issues of project management. In that respect, it was through a focus on deadlines that my general interest for time in projects and organizations began. This chapter is my attempt to summarize what I have learned about deadlines over the last 20 years or so. It is also an opportunity to summarize a few managerial insights that I believe are important to make deadlines work in a better way – what you need to think of to make deadlines work.

DOI: 10.1201/9781003502654-10

The approach taken here is that deadlines may inform us about the very basic question of why projects exist – and what the role of project management really is (Söderlund, 2004). Deadlines may also inform us about the very essence of projects and how action is triggered in the context of projects (Shenhar and Dvir, 2007). This seems particularly important because action is normally considered to be the central idea of project organizing (Lundin and Söderholm, 1995) – to make things happen, to trigger collaboration, to foster transformation and to drive creativity. These are all central ideas of managing projects. In that respect, I will argue that deadlines are critical for understanding projects as a particular form of organizing and for informing us about some of the very basic ideas about project management. Perhaps this may sound a bit philosophical and to some extent it really is, but I will also try to keep it as concrete as possible and explore how managers can use deadlines in a better way.

A WORD ON DEATH

The word deadline is used frequently in various project contexts: in event projects, in media projects, in marketing projects, in product development projects, etc. The deadline stipulates a point in time when something is expected to be finished; it is a time frame, a time limit. In many contexts, it is more than that – it stipulates when something must be completed. Many would argue that the most important success criterion is in fact whether a project meets its deadline or not (see for instance Brown and Eisenhardt, 1997). Late projects mean big problems, sometimes. Late projects mean complete failure, sometimes, as would be the case in many event or sports projects. That said, some projects just don't seem to be able to finish on time, and others just go on forever almost like they do not want to finish (Flyvbjerg, 2014).

The point here is that deadlines need to be understood, they need to be managed and they need to be used to make project management work. A project without a deadline is perhaps not a project. A project manager without an understanding of deadlines and how to use them, is considerably constrained. However, our research has indicated that far too many practitioners misunderstand the power of deadlines and fails to see that it is a managerial concern – not always something that is given from above, but in fact controlled by management. Deadlines are managerial instruments that need to be taken seriously.

The word deadline was taken from the civil war when a line was drawn to indicate where prisoners could not pass without the consequence of being shot by the guard. It clearly signaled a consequence – and in that respect, it told everyone involved that crossing the deadline meant death (Gersick, 1995). In a project context, there are consequences. In some cases, these consequences are severe. Very seldom the consequence is death, although there might very well be cases like that also in a project context, for instance when it comes to rescue operations or the development of new drugs. However, an important idea of organizing by projects is to make people aware of the consequence of not meeting the deadline – this might be loss of a business opportunity, lack of technological progress, or missed societal improvements. Think of the construction company delivering a new building for an airport, like the famous case of Terminal 5 at Heathrow (Davies et al., 2016). Not meeting the deadline may have severe consequences, might lead to substantial penalties, but more importantly might lead to the fact that the entire airport will not be operational. The loss for the client could be substantial. In some cases, these consequences might be managerially constructed, not necessarily client-centered. For instance, that tickets have been sold, that stakeholders have been invited to examine a prototype of a new product, that the production of the old automotive model has been phased out to give room for the new model. Not meeting the deadline may cost a fortune, in fact lead to bankruptcy for the company in charge. At least, not meeting the deadline might be embarrassing to begin with, and lead to lack of trust among suppliers and clients. Sticking to the deadline is important for a number of reasons, for gaining trust, for allowing others to pace themselves in accordance with the deadline.

WHY PROJECTS EXIST

When initiating a project the deadline constitutes one of the most important rationales; one of the most important ambitions with the project. It is also often stated that deadlines need to be set quite strictly. The so-called "Parkinson's law" informs us that available time in a project is always consumed. This principle was introduced by Cyril Northcote Parkinson in an article in 1955 in the Economist and later addressed in a book (Parkinson, 1955). This law builds on the observation that the duration of public administration expands to fill its allotted time span, regardless of the actual amount of work to be done.

This principle underlines the centrality of setting deadlines sensibly. In fact, it suggests that people should not have too much time to be able stay effective, they will get involved in activities and tasks and make work that are not important to the project, not important for completing what must be done. In fact, some of these added tasks and extra work may in fact endanger the tasks that are critical for the project.

This is a very important aspect of a deadline – deadlines should be tough, they should be challenging, but not impossible to meet. People need to believe that they have a chance making it. The deadline should trigger some kind of rethinking with regard to how the task is going to be solved. In that regard, it is important that deadlines are challenging to make people rethinking their ordinary way of doing things. If projects should be triggering rethinking then deadlines should be set accordingly – to make people reflect about new ways of doing things (Obstfeldt, 2017).

Thus, the existence of a deadline is fundamental to most projects. In that regard, deadlines give the organization and the people who are working on the project a "time giver" – a sense of direction of what is essential in the project. This is the first purpose of the deadline – as a form of pacing instrument with the external environment, be that the client (delivery of a new power plant) or the operations organization (like production start of a new automobile). The time giver could be any central cycle, event, or date in the environment and industry that is of essence to the output of the project. This would then make project participants more aware of why the deadline is important and how the project is connected with its environment – from a time perspective. This time adjustment activates a number of processes that are important for the project to relate to its environment – it makes people think about what needs to be done, it makes people think in terms of what must be done rather than what could be done. In that regard, deadlines serve a rather particular kind of goal orientation. It tells people what they must do. We tend to speak about this as the cooperation function of deadlines – we are in this together and this is what we must do.

This observation is an important part of why we organize activities as projects – why projects exist (Söderlund, 2004). It clearly points out that what we refer to as "time pacing" is an important part of organizing by projects. In that regard, the deadline indicates that we need to pace our activities with the external environment. Not responding to the external requirements with regards to the deadline will lead to network externalities – clients waiting for a system not delivered, airport operators not able to run the airport, or educators being idle waiting for an IT system that is not operational. This is the first point about deadlines. This point underlines that deadlines trigger project partici-pants to make calculations in terms of the ambition of the project, in terms of what is possible to do within the available time. In that regard, the deadline produces some kind of "rationalistic break" – it requires people to think about what they should be doing, and what they shouldn't be doing. This is an important part of why we organize work as projects – to make people more aware of what is important to reach the overall goal.

THE LOGIC OF CONSEQUENTIALITY

In organization theory, there are two logics of organizing that are particularly important and noteworthy. The first one relates to the so-called "logic of appropriateness" (March and Olsen, 1976). Here we have rules, standard-operating procedures, role descriptions, and instructions. I do what I do because this is what is expected from me, this is what is stipulated in my job description. This is a very common logic in most organizations, in bureaucracies, in repetitive organizing. However, deadlines are meant to get people to break out of their logic of appropriateness and instead focus more on what is commonly referred to as the "logic of consequentiality." This organizing logic is supposed to make people more focused on the consequences of doing certain things, not according to particular descriptions or expectations but more oriented towards goals, ambitions, and set targets. In that regard, it is a logic that operates quite differently from the logic of appropriateness.

Deadlines are to a much greater extent a way of controlling behavior following the logic of consequentiality. In that respect, one might argue that the entire idea of organizing by projects is to get people to think about consequences, to think about what must be done, to think about how to make the deadline, to get things done, and to stimulate action. It is a rather different control mechanism then than the mechanisms that are typically associated with the logic of appropriateness.

In a large-scale systems development project that we studied for several years – a challenging project within global telecom giant Ericsson, the project manager talked about the importance of "lagom" – a Swedish term for making things "just right." The project needed to be finished on time – that was the most important thing. The delivered system had to contain the most important features and operate as promised to the client – and it had to be delivered on time. To make this happen, the project manager, underlined the importance that the engineers working on the project needed to focus on what was critical to the project – "what must be done" – on what was the essence of the project. Throughout the entire project, the overall deadline played a very central role to ensure that the project was on track and on time. The project managers used the overall deadline to plan key testing activities, to estimate when major deliveries had to be done, which then also became a force for action within the project – the deadline also made people to feel strongly about when things had to be done and that these time limits had to be respected. This principle of a time-centered project management approach in which overall deadlines and milestones along the way later on led the firm to speak of an "integration-driven project management model" (Lindkvist et al., 1998) in which overall deadlines played a very central role. The milestones implemented along the way became a way of implementing the overall deadline by building the system step-by-step by using collective integration arenas and tests involving several teams during the entire project.

MINDFUL ABOUT TIME

The second point relates to how deadlines are used internally and how it might make project participants more engaged in other participants' time orientations and time horizons. After project participants have begun engaging in an analysis of what needs to be done – and how the deadline links with other activities in the project environment, then project participants will move into a process where they become more aware of how their activities are related to other participants' activities (Söderlund and Pemsel, 2022). In that regard, the deadline makes project participants more aware of the other participants' time orientations and problem-solving cycles. In that regard, they will become more responsible for their part in coordinating the project, in relating their activities with other activities within the project. We tend to refer to this as the coordination function of deadlines.

In short, deadlines should make people more aware of the importance of timing, of sequences, of durations, and of the speed of the project. In that respect, the deadline has an important coordination role to play. From these two key steps (externally and internally) of generating activities from the deadline, project participants should then also be able to plan key activities keeping the deadline in mind. For instance, the opening day is June 1, the general rehearsal needs to be at least one month prior to that – and six weeks prior to the general rehearsal, there is a need to set the costumes and try out the equipment needed for the performance, and four weeks prior to that we need have the choreography in place, and so on. For doing that, deadlines need to be non-negotiable. To set the rhythm, to set the pace, to create the momentum to which everyone involved need to respond.

DEADLINES FOR COOPERATION AND COORDINATION

The deadline is set – it should make people objectively communicate when activities should be done. In that regard, project organizing is a performance art – deadlines should be public and people should be aware of when the deadline is. More so, for those involved in the project, knowing when the deadline is serves an important ground for establishing effective collaboration – it is an instrument both for cooperation (agreeing on what and when things should) and an instrument for coordination (to relate one's own activities with the others' activities). Within the project management community, the latter has been referred to as an issue of "knowledge entrainment" (Söderlund, 2010). Research on deadlines and projects has made considerable progress the last couple of decades. In fact, the notion of deadlines has been considered one of the most essential features of projects as organizational forms. Lindkvist et al. (1998) in a much-cited paper pointed out that deadlines are a particular kind of control form, which is implemented through various sorts of milestones and problem-solving arenas during the implementation of the project. Research has since suggested that there are a few particularly important aspects of deadlines that one needs to consider either when implementing or managing them.

These could be formulated in the following way by answering the overall question of what is important to make deadlines work:

- Deadlines need to be non-negotiable. The date should be specified and clear to everyone involved.
- Deadlines should be public.
- Deadlines need to be communicated to everyone involved on a regular basis.
- Deadlines must have consequences – deadlines should be deadlines.
- Deadlines need to be challenging, yet realistic.
- Deadlines need to produce a so-called rationalistic break and make people more aware of time in general.
- Deadlines need to get people into adopting a logic of consequentiality and make people think through what must be done to meet the deadline.
- Deadlines need to be linked to external "time pacers," i.e., various time givers and key activities in the environment that dictate when things need to be done, for instance seasons, industry events, and industry cycles.
- Deadlines need to make people more aware of overall time pacers and also others' temporal orientations and priorities.
- Deadlines need to make people more aware of other people's time – make them more concerned about the pace and rhythm of other stakeholders and project participants.
- Deadlines need to be implemented through milestones and problem-solving arenas during the life of the project. These arenas make people aware of the fact that their deadline matters to other people involved in the project – and that the consequences of a missed deadline are real.
- Deadlines need to be public, people should know about the deadline, and managers may use public events, gatherings, open meetings to come together and identify if deadlines have been met.

So, the idea with this chapter was to give the reader an idea of the power of deadlines, of the importance of deadlines, of how deadlines serve important roles for establishing both cooperation and coordination in a project. To the practitioner, I would like to say that you need treat deadlines as a powerful instrument in your project, be careful when you set or accept a deadline, communicate the deadline, speak about the deadline, make people aware of the deadline and use it to create a sense of meaning and commitment. Use the deadline to make people aware of time and timing in general and that the deadline make people in the project more mindful about time in the project and other people's time, for instance, how long people need to prepare a certain activity, how long it takes to complete a particular task in the project. Breathe life into your project by using the deadline. The awareness of death should make people more careful about life in the project.

REFERENCES

Brown, S. L. & K. Eisenhardt (1997). The art of continuous change: Linking complexity theory and time-paced evolution in relentlessly shifting organizations. Administrative Science Quarterly, 42: 1–34.

Davies, A., M. Dodgson & D. Gann (2016). Dynamic capabilities in complex projects. The case of London Heathrow Terminal 5. Project Management Journal, 45(2): 26–46.

Flyvbjerg, B. (2014). What you should know about mega-projects and why. Project Management Journal, 45(2): 6–19.

Gersick, C. J. G. (1995). Everything new under the gun creativity and deadlines. In Creative action in organizations. C. M. Ford and D. A. Gioia (eds.), 142–148. Thousand Oaks: Sage.

Lindkvist, L., J. Söderlund & F. Tell (1998). Managing product development projects: On the significance of fountains and deadlines. Organization Studies, 19(6): 931–951.

Lundin, R. A. and A. Söderholm (1995). A theory of the temporary organization. Scandinavian Journal of Management, 11(4): 437–455.

March, J. G. & J. P. Olsen (1976) Attention and the ambiguity of self-interest. In Ambiguity and choice in organizations. J. G. March and J. P. Olsen (eds.), 38–53. Oslo: Scandinavian University Press.

Obstfeldt, D. (2017). Getting new things done: Networks, brokerage and the assembly of innovative action. Stanford: Stanford University Press.

Parkinson, C. N. (1955). Parkinson's law. London: The Economist.

Shenhar A. & D. Dvir (2007). Reinventing project management: The diamond approach to successful growth and innovation. Boston: Harvard Business School Press.

Söderlund, J. (2004). Building theories of project management: Past research, questions for the future. International Journal of Project Management, 22: 183–191.

Söderlund, J. (2010). Knowledge entrainment and project management: The case of large-scale transformation projects. International Journal of Project Management, 28(2): 130–141.

Söderlund, J. & S. Pemsel (2022). Changing times for digitalization: The multiple roles of temporal shifts to enable organizational change. Human Relations. 75(5): 871–902.

11

Our Love Affair with Project Slack (and Why It Ruins Our Schedule Accuracy)[1]

Jeffrey K. Pinto[1] and Kate Davis[2]
[1]Black School of Business, Penn State, Erie, Pennsylvania, United States
[2]Cranfield School of Management, Cranfield, United Kingdom

In this chapter, we elected to tackle a thorny problem with the manner in which project plans are created. Many of us are quite familiar with the steps for preparing a project schedule in theory, following the best practices within our firms, through to guidance from professional bodies of knowledge, or from reading textbooks. Theory and best practice suggest an orderly series of steps, including: (1) activity identification, (2) activity duration estimation using either deterministic or probabilistic methods, (3) precedence diagramming, and (4) forward and backward passes through the network to identify expected duration and activity slack (Pinto, 2019). The result, we have been taught, is the creation of an efficient, logical diagram of project development, an accurate schedule to completion, and a series of interactive precedent and successor activities that allow us to accurately monitor project execution. If only it were truly this easy!

The real problem, however, is not with the "science or theory" of scheduling, but how the process translates into real behavior in practice. Specifically, we often find that the law of unintended consequences frustrates our best intentions in project planning (Pinto, 2014) and scheduling practices, often as a result of mixed messages, perverse incentives, and misguided project team member behaviors aimed more at self-preservation than project efficiency. This chapter is intended to take an "eyes-open" look at the manner in which our best efforts to

DOI: 10.1201/9781003502654-11

develop accurate project schedules are often frustrated by a variety of human behavior mistakes that are recurring and, in fact, predictable. In identifying these mistakes, it is important to point out a couple of qualifying points: first, we are not suggesting that *all* these behaviors occur *all* the time on the part of *all* members of a project team. That's too easy and invariably, it is inaccurate. Some of us are more prone to certain of these mistakes and indeed, some organizations and their cultures tend to promote self-protecting behavior (also known as "CYA," or, *Cover your Ass*) to greater degrees than other firms. The key here is to take an honest, open-eyed look at the pressures that "could" occur and weigh them in the crucible of your own experiences. If you find evidence of some of these issues occurring in your past experiences, it suggests that you can reflect on the reasons why this may be the case and more importantly, take remedial steps to avoid them in the future.

This chapter will examine the manner in which we create our activity schedules. As we suggested, the science of logical steps is clear and *should* result in a clear activity schedule to the finish line. The critical miscues that we are focusing on here are those concerned with the way we create the schedule and determine activity durations, both of which are highly important for managing the project. Unfortunately, this result rarely occurs; projects are messy, prone to error, and affected by a variety of human decisions and behavioral miscues (Pinto, 2013). To put it another way, if we don't create a sensible plan at the start, all our management downstream won't amount to much. So, with this idea in mind, let us consider some of the most common ways that people can mess up perfectly good project schedules, recognizing that the root causes revolve around a perspective of human behavior at work, miscommunication among teams and their managers, and our natural desire to minimize personal risk.

REASON 1: BUILDING IN PERSONAL SAFETY

The initial step in building an accurate project schedule consists of first disaggregating projects into their incremental activities through a concept known as the work breakdown structure (WBS). In addition to identifying all necessary tasks to complete the project, this early planning stage also requires the project manager and team to prepare reasonable duration forecasts for each activity in the project. The "wild card," of course, is the individual making the estimate: how confident is this person in their calculations and how much incentive do they have to provide an accurate estimate, as opposed to one that is padded to provide a degree of safety?

Goldratt (2017) and others have argued that a variety of human behavioral elements factor heavily into the manner by which individual project team members come up with their activity duration estimates; most particularly, the need for self-protection against the risk of sanctions in the cases of missing agreed-upon target dates ("What happens if I promise to be done in three days and it takes me five?"). In other words, the personal and professional risks associated with promising a time to completion (X) but ultimately delivering the completed task at some later point (X + Y) means that project team members are motivated to "pad" their estimates with some personal safety to account for unexpected delays, technical problems, demands from multitasking obligations, and other events that can delay their completion of assigned work (Umble et al., 2006). When asked for a reasonable time estimate probability for successful completion of their activities ("How long will this take you to complete?"), team members are naturally loathed to risk their reputation on a "maybe" estimate, preferring to give time estimates that offer reasonable assurance of targets they are capable of hitting. It is not surprising, then, that individuals would seek to gain for themselves a margin of safety against any potential surprises or unanticipated events that could delay them.

Activity completion rarely neatly follows an estimated time schedule,, the reality is that due to a variety of complications and challenges (many unforeseen), project tasks may be completed a bit earlier than estimated, but they also have the capacity to be completed much, much later. In short, a perfectly reasonable initial task completion estimate can blow up through our own mistakes, or forces outside of our control. Thus, in Goldratt's (2017) words, when project team members are asked to give estimates of their activity completion time (and with the desire to be reasonably certain that the estimate they give is one they can actually hit), these team members would much prefer aiming for 90% certainty as opposed to only a 50/50 likelihood. Critically, when employing the log-normal distribution, this desire for achieving a 90% likelihood of success leads team members to propose time estimates that can inflate their initial estimate by over 200% (Goldratt, 2017)! It is hard to miss a deadline that contains a 200% inflated delivery date.

REASON 2: THE NEW MATH, OR 2 + 2 + 2 = 8

Following the logic of our first cause for lengthening the project, the desire to build in personal safety, we can now apply that same motivation up the ladder, or for managers of multiple project team members. For bottom-up estimating, the individual task time estimates submitted by each project team member travel up to next-level supervisors—either project managers or their subordinates, as each activity duration estimate is collected to develop the aggregate estimates for the project (Leach, 2003). Each supervisor—responsible for the performance of their element of the project and their team members— faces their own concerns about accuracy of the final estimates and conforming to a reasonable projection of project completion. Moreover, the team manager may harbor an implicit suspicion that the estimates provided by each member of their team are not sufficiently robust to ensure the manager with their own degree of safety, as they recognize that the performance of their team in completing a part of the project will reflect on themselves and their relation with top management (Graham, 1994). So, just as with each individual team member, who have already done their own calculation of the need to adjust estimates to reflect a greater degree of certainty in completion (the 90% probability, or 200% estimate increase), aggregated estimates contain significant padding as the supervisor often adds in their own measure of safety on top of the aggregated estimates. In this way, a manager of three subordinates, with each estimating that their activity will take two days to complete, aggregates a "revised" total of, not six days, but a more cautious eight days to account for the supervisor's own need for safety. Do you see the irony? In "protecting" themselves and their department, these supervisors are adding their own safety margin *on top of already inflated estimates*. And so, inflated individual estimates become doubly (or triply) padded, as they travel further up the managerial ladder.

REASON 3: ANTICIPATING TOP MANAGEMENT CUTS

In his book, *Project Management as if People Mattered*, Robert Graham (1989) noted a commonly observed phenomenon regarding less-than-authentic interactions between project teams (often including the project manager) and their top management, motivated to realize the completion of the project as quickly as possible; that is, top management schedule intervention in the form of post hoc cuts. Once a project schedule has been assembled and the resulting network created, activities linked, and a critical path developed, it is not uncommon at this point to receive feedback from key members of top management summarized with a simple response: "It's too long; shorten it." Regardless of their reasons (some perfectly legitimate) for these peremptory commands, project managers and team members quickly pick up on this ploy, recognizing the motivation driving it (or the managers most known for pushing these shorter windows).

Managers who get a reputation for shaving a percentage of time in activity estimates encourage an equally predictable reaction on the part of their subordinates, who are motivated to add in extra time to account for these expected cuts. For example, Manager A may have a reputation for shaving 15% off initial project schedules, prompting his subordinates to anticipate these arbitrary cuts by simply adding a 15% safety margin to the schedule they present to the boss. Thus, one knee-jerk response motivates a second one, as a cycle of inauthentic behaviors manifests, with the real goal here one of self-preservation, rather than mutually working to find legitimate ways to shrink a project schedule. As Graham (1994) was fond of saying, these exchanges can be summed up in a simple statement made by generations of frustrated project team members: "If you don't take my estimates seriously, I'm not going to give you serious estimates."

Based on the above arguments regarding behavioral factors in project duration estimates and the steady inflating of these estimates to account for factors outside of their control, it would be reasonable to support our original contention; namely, that no project should ever be late. With the individual safety, manager's safety, and additional padding due to anticipated trimming from top management, it is easy to assume that all these bloated schedules should guarantee that no project will ever over-run its schedule. Unfortunately, we know that this is not the case; project after project in diverse industries continue to run over schedule, with some delays running into additional months, or even years. How, then, do we account for these delays when so much effort is taken by so many organizational members to protect themselves by adding levels of safety to their estimates? The simple answer, is that just as project managers and team members are adept at finding ways to "enhance" their schedules during duration estimation, they prove equally adroit (intentionally or through patterned mistakes) at squandering these advantages during project execution (Englund and Graham, 2019). Thus, we do an excellent job of protecting ourselves during the planning process and then throw much of this safety away during actual project development.

NOTE

1 Portions of this chapter were adapted from: Pinto, J.K. (2022), "No project should ever finish late (and why yours probably will, anyway)", *IEEE Engineering Management Review*, 50, 181–192.

REFERENCES

Englund R. & Graham R. J. (2019). *Creating an Environment for Successful Projects*. San Francisco, CA, USA: Berrett-Koehler Publishers.

Goldratt E. M. (2017). *Critical Chain*. New York, NY, USA: Routledge. Graham R. (1994). Personal communication.

Graham R. (1989). *Project Management as if People Mattered*. Bala Cynwyd, PA, USA: Primavera Press.

Graham R. (1994). Personal communication.

Leach L. P. (2003). Schedule and cost buffer sizing: How to account for the bias between project performance and your model. *Project Management Journal*, 34(2), 34–47.

Pinto J. K. (2013). Lies, damned lies, and project plans: Recurring human errors that can ruin the project planning process. *Business Horizons*, 56, 643–653.

Pinto J. K. (2014). Project management, governance, and the normalization of deviance. *International Journal of Project Management*, 32, 376–387.

Pinto J. K. (2019). *Project Management: Achieving Competitive Advantage, 5th Ed*. Upper Saddle River, NJ: Pearson.

Umble M., Umble E., & Murakami S. (2006). Implementing theory of constraints in a traditional Japanese manufacturing environment: The case of Hitachi Tool Engineering. *International Journal of Production Research*, 44, 1863–1880.

12

How Principles Can Make Agile Benefits Realisation Successful

Carl Marnewick
University of Johannesburg, Johannesburg, South Africa

Project management is very much focused on processes. This has been the case since the first best practices were documented in standards such as the PMI's PMBoK® Guide. The 6th edition of this standard has 49 processes spread over 10 knowledge areas. The same applies to other project management standards and methodologies such as PRINCE2 and the APM Body of Knowledge. Yet, in spite of all these processes, projects are still not delivering on the value or benefits envisaged. This is irrespective of the industry or size of the project. The IT industry is no exception to this trend and anecdotal evidence (Chaos Chronicles and Prosperus Reports) highlights that IT projects are not realising benefits in approximately 30% of the cases. This is a trend that stretches over decades internationally as well as where I live in South Africa. These results are largely based on IT projects adhering to a traditional approach like Waterfall.

This prompted the introduction of the Agile Manifesto in 2001. The Agile Manifesto[1] is based upon 12 principles that guides the development of software. The focus is on agility and not so much on processes. Various studies highlighted that IT projects following an agile approach are more prone to realise benefits than IT projects following a traditional approach. Although there is evidence Agile projects are more successful than traditional projects, that might not be enough to convince sceptical senior managers. To realise even more benefits, there are key principles that need to be in place. These principles are the spring board for improving benefits realisation. This shift from processes to principles is

DOI: 10.1201/9781003502654-12

also adopted by PMI as the 7th edition of the PMBoK® Guide introduces 12 principles and eight performance domains. The adoption of principles can play an important role in putting the appropriate Agile benefits realisation processes in place. In other words, the adopted Agile principles will dictate the Agile processes and practices.

A PROCESS VIEW OF BENEFITS REALISATION

Benefits are associated with the change that the product or service of the project brings about. The focus is not in the product itself but the change. An example is a mobile phone. The benefits associated with a mobile phone is the ease of communication associated with it and access to mobile banking to name but a few. The more benefits associated with a product, the better. Benefits realisation or benefits realisation management is the process of managing the benefits associated with the project's deliverable. The rationale of benefits realisation is to maximise value of the deliverable.

Over the years, various authors highlighted the process of benefits realisation. Typically, these processes consist of identifying the benefits, executing the benefits, manage the benefits and sustaining the benefits. Yet, in spite of all these efforts, benefits are not realised or harvested. The PMI estimates the percentage of organisations that realise benefits at between 56% and 78%. The question that needs to be asked is why benefits are not realised when there are documented processes. Are processes simply not followed? Are organisations' profits so high that they don't care? Since a process approach to benefits realisation is not delivering the anticipated benefits, an alternative approach is required. This alternative comes in the form of principles.

PRINCIPLES FOR AGILE BENEFITS REALISATION

The focus of principles is on the outcomes (benefits) rather than the process or deliverables. The nice thing about principles is that they are not prescriptive at all. Agile project teams adopt principles to guide their behaviour of how they want to realise benefits. Which, how, and when the principles are adopted is totally up to the agile project team. By adopting the Agile benefits realisation principles, the agile project team changes the focus away from rules and processes to a more ethical decision-making where decisions are made based on what is best for the organisation at large as well as for the project; And not just the adherence to rules and processes that are meant for a specific situation.

Twelve agile benefits realisation principles have been identified. These 12 principles can be divided into three groups: must-have, essential, and technical principles. Some of these principles are specific to an agile environment but can be adopted to non-agile environments as well.

The three must-have principles are fundamental to agile benefits realisation and are as follows:

- *Consequence management:* This is the most important principle and has two features. The first feature is that decisions do have consequences. Team members are accountable for their decisions. The second feature is that the outcome of a decision has a wide influence on the organisation, the project and benefits realisation. When decisions are made, all consequences need to be evaluated to determine the positive and negative impacts of the decision. No one is exempted from the consequences of their decisions.
- *Visibility:* This principle goes together with consequence management. When work is visible, it creates an environment of trust. Visibility showcases current projects and its associated benefits to everyone. This openness and visibility are achieved through Kanban boards and Enterprise Visibility Rooms (EVRs). When decisions and subsequent consequences are visible, team members are more conscientious in the way they work.
- *Consistency:* There are many ceremonies in an agile project environment. Ceremonies are events that happen at the same time, for instance, daily-standups that happen every day at a specific time. Or a sprint that occurs that stretches over two weeks and follows a specific process. When it comes to benefits realisation, program increments (PIs) are important ceremonies. PIs are the periods when work is delivered, benefits are determined, and fast feedback is received. Consistency in PIs provides consistent results.

The following seven principles are essential to agile benefits realisation but some might be optional such as Beyond budgeting.

- *Beyond budgeting*[2]: This is a new concept and is based on organisational agility. Beyond budgeting improves performance using flexible sense-and-respond mechanisms and consists of 12 principles. These principles are divided into leadership and management. The leadership principles are (i) purpose, (ii) values, (iii) transparency, (iv) organisation, (v) autonomy and (vi) customers. The management principles are (i) rhythm, (ii) targets, (iii) plans and forecasts, (iv) resource allocation, (v) resource allocation, and (vi) rewards.

- *Course correction:* When a process-approach of benefits realisation is followed, misalignment is only realised once the final product is delivered. The consequences can then be devastating. The principle of *course correction* forces the agile team to make rapid changes when they realise that the intended benefits are not going to be realised. This rapid change typically occurs during a PI session.

- *Define target outcomes:* This principle is a universal principle. Projects and their associated benefits must have well defined target outcomes. These outcomes are derived from the organisation's vision and strategies. This ensures alignment between the organisation's vision, the project's vision, and the realised benefits.

- *Behavioural change:* Just as agile is mindset, so is benefits realisation. Agile team members need to consistently think about the benefits of the product and make continuous *course corrections* (when required). This requires a change in behaviour as the focus is not on the process and ticking boxes but on the ultimate reason for the product and the benefits that it should realise.

- *Customer journeys:* Customer journeys indicate the various touch-points that the customer have with the organisation. This is typically presented in a map. The agile project team that develops a product or customises an existing product, needs to understand the customer journey. They need to map and link the benefits to the customer journey. This allows the agile project team to have a holistic view of when the product will enhance the customer's experience through the realisation of benefits.

- *Continuous monitoring:* This principle is closely associated with the *Course correction* principle. Agile allows for continuous monitoring through sprints and PI's. At the team level, the team can make course corrections after each sprint which is typically every two weeks. At the end of the sprint, a retrospective (which is a ceremony and relates to *consistency*) is held. The retrospective creates an opportunity for the agile project team to inspect itself and create a plan for improvements to be enacted during the next sprint. At a programme level, PI's are used for continuous monitoring. This typically happens once a quarter and is spread over 12 sprints.

- ***Cross-functional teams:*** These teams incorporate business and technical people. The rationale behind cross-functional teams is that that these teams make their own decisions. These decisions can be business-related or technical related. It reduces time as no escalations are required. Since business is part of the cross-functional team, decisions that influence benefits realisation can be quickly dealt with. This is highly encouraged in Agile. Cross-functional teams are more connected especially across different departments. Team members that feel connected trust one another, are better at problem-solving and conflict resolution.

The following two principles are of a more technical nature. Although these principles are not directly linked with benefits realisation, it creates the platform from which products are launched. If the technical platform is not optimally configured, the launch of products is compromised.

- ***Build-own-run:*** This principle rests on the philosophy that teams take ownership for their own work. Teams take ownership of the entire product from development right through to production. A sense of ownership and commitment is instilled. This principle is related to the principles of *visibility* and *consequence management.*
- ***Architectural runway:*** This principle focuses on existing code, components and technical infrastructure that are required to support the realisation of benefits. The architectural runway provides the technical foundation for cross-functional teams to *build-own-run* the product.

SUMMARY

Two principles, *visibility* and *consequence management* are non-negotiable and should be the foundation upon which agile benefits realisation is built. *Consistency* is also an important principle and the focus should be on the consistency of performing ceremonies.

Consistency is key in a principle-based environment as behaviour should always be the same, irrespective of the environment. This is applicable to decision-making and performing of ceremonies. The implementation of the remainder of the principles is left to the discretion of the agile project team. One such principle is *beyond budgeting*. Not all organisations are ready to go this route and thus it can be excluded until required.

By no means, are these 12 principles all the principles. More principles will manifest as time goes on and practitioners apply these principles. But these principles can help build support for Agile benefits realisation and keep it strong within the organization.

NOTES

1 https://agilemanifesto.org
2 https://bbrt.org

REFERENCES

Ashurst, C. (2011). *Benefits Realization from Information Technology*. Palgrave Macmillan.

Bradley, B. (2010). *Benefit Realisation Management: A Practical Guide to Achieving Benefits Through Change* (2 ed.). Surrey, England: Gower Publishing Limited.

Khoza, L., & Marnewick, C. (2020). Waterfall and Agile information system project success rates – A South African perspective. *South African Computer Journal, 32*(1), 43–73. doi:10.18489/sacj.v32i1.683

Labuschagne, L., & Marnewick, C. (2009). *The Prosperus Report 2008: IT Project Management Maturity vs. Project Success in South Africa*. Johannesburg: Project Management South Africa.

Marnewick, C. (Ed.) (2013). *Prosperus Report – The African Edition*. Johannesburg: Project Management South Africa.

Marnewick, C., & Marnewick, A. L. (2022). Benefits realisation in an agile environment. *International Journal of Project Management, 40*(4), 454–465. doi:10.1016/j.ijproman.2022.04.005

Petit, Y., & Marnewick, C. (2023). Fixed Capacity and Beyond Budgeting: A Symbiotic Relationship within a Scaled Agile Environment. In V. Anantatmula & C. Iyyunni (Eds.), *Research Handbook on Project Performance* (pp. 198–215). Cheltemham, United Kingdom: Edward Elgar Publishing Ltd.

Project Management Institute. (2017). *A Guide to the Project Management Body of Knowledge (PMBOK® Guide)* (6 ed.). Newtown Square, Pennsylvania: Project Management Institute.

Project Management Institute. (2019). *Benefits Realization Management: A Practice Guide*. Newtown Square, Pennsylvania: Project Management Institute.

Project Management Institute. (2021). *A Guide to the Project Management Body of Knowledge (PMBOK® Guide)* (7 ed.). Newtown Square, Pennsylvania: Project Management Institute.

Serrador, P., & Pinto, J. K. (2015). Does Agile work? — A quantitative analysis of agile project success. *International Journal of Project Management*, 33(5), 1040–1051. doi:10.1016/j.ijproman.2015.01.006

The Standish Group. (2014). *Chaos Manifesto 2014*: The Standish Group International.

The Standish Group. (2018). *The CHAOS Report: Decision Latency Theory*. Retrieved from https://www.standishgroup.com/store/services/10-chaos-report-decision-latency-theory-2018-package.html

Ward, J., & Daniel, E. (2012). *Benefits Management: How to Increase the Business Value of Your IT Projects* (2 ed.). United Kingdom: John Wiley & Sons.

13

The Relationship Between Project Success and Project Efficiency[1]

Pedro M. Serrador
Northeastern University, Toronto, Ontario, Canada

"Eighty percent of success is showing up."

– Woody Allen

Project success criteria have been measured in a variety of ways. While the conventional measurement of project success has focused on tangibles, current thinking is that ultimately, project success is best judged by the stakeholders, especially the primary sponsor, (Turner and Zolin, 2012). As Turner and Zolin (2012) note, assessing success is time-dependent: "As time goes by, it matters less whether the project has met its resource constraints; in most cases, after about one year it is completely irrelevant. In contrast, after project completion the second dimension, impact on the customer and customer satisfaction, becomes more relevant." (Shenhar et al., 1997, p. 12). In a similar vein, Cooke-Davies (2002) differentiated between project success and project management success. Project management success is the traditional measure of project success, measured at project completion, and is primarily whether the output is delivered to time, cost, and functionality (the so-called triple constraint). Following Shenhar and Dvir (2007), we called this project efficiency. Project success is whether the project outcome meets the strategic objectives of the investing organization. In this paper we focus on overall project success which is measured by how satisfied key stakeholders are about how well the project achieves its strategic objectives.

Munns and Bjeirmi (1996) noted that much of the project management literature considers "projects end when they are delivered to the

DOI: 10.1201/9781003502654-13

TABLE 13.1

The Five Dimensions of Project Success After Shenhar and Dvir (2007)

Success Dimension	Measures
1. Project efficiency	Meeting schedule goal
	Meeting budget goal
2. Team satisfaction	Team morale
	Skill development
	Team member growth
	Team member retention
3. Impact on the customer	Meeting functional performance
	Meeting technical specifications
	Fulfilling customer needs
	Solving a customer's problem
	The customer is using the product
	Customer satisfaction
4. Business success	Commercial success
	Creating a large market share
5. Preparing for the future	Creating a new market
	Creating a new product line
	Developing a new technology

customer," (p. 83). This focus on the end date of the project is understandable from a project and project manager's standpoint. The definitions of a project imply an end date; at that time the project manager is likely to be released or move on to another project. Also the reward structure in many organizations encourages the project manager to finish the project on cost and time and little else, (Turner, 2014) (Tables 13.1–13.3).

The literature has also examined the wider impact of projects on the business. Customer satisfaction has long been a part of the project management literature (Kerzner, 2009), but it has not usually been included in the formal measures of success. Researchers increasingly measure success by impact on the organization rather than just meeting the triple constraint. Dvir, Raz, and Shenhar (2003) state that "there are many cases where projects are executed as planned, on time, on budget and achieve the planned performance goals, but turn out to be complete failures because they failed to produce actual benefits to the customer or adequate revenue and profit for the performing organization." (p. 89). They also found, "all four success-measures (meeting planning goals;

TABLE 13.2

Project Success Rating for All Projects

	Valid N
Failure	98
Not fully successful	259
Mixed	345
Successful	451
Very successful	233
All groups	1,386

end-user benefits; contractor benefits; and overall project success) are highly inter-correlated, implying that projects perceived to be successful are successful for all their stakeholders." (p. 94). Thomas, Jacques, Adams, and Kihneman-Woote (2008) state that measuring project success in not straightforward. "Examples abound where the original objectives of the project are not met, but the client was highly satisfied. There are other examples where the initial project objectives were met, but the client was quite unhappy with the results." (p. 106). Collyer and Warren (2009) cite the movie Titanic, which was touted as a late, over budget flop but went on to be the first film to generate more than $1 billion. The Sydney Opera house was completed ten years behind schedule and 1,357% over budget. In the end, it became a symbol not only of Sydney but of Australia. Now hailed as "an architectural marvel" and one of "the most influential places in history," the Sydney Opera House is the youngest site to achieve UNESCO World Heritage status.

We define the two competing measures as:

Project efficiency: meeting cost, time and scope goals.

Project success: meeting wider business and enterprise goals as defined by key stakeholders.

Few people have investigated to what extent these two measures of success are correlated. Prabhakar (2008) notes "There is also a general agreement that although schedule and budget performance alone are considered inadequate as measures of project success, they are still important components of the overall construct. Quality is intertwined with issues of technical performance, specifications, and achievement of functional objectives and it is achievement against these criteria that will be most subject to variation in perception by multiple project stake-holders." (p. 7). But, as we saw above, the components of project efficiency are neither necessary nor sufficient conditions of success. So what, if anything, is the relationship between project efficiency and project success? This leads to our research question:

To what extent is project efficiency correlated with overall project success?

A survey was undertaken which gathered responses from a total of 859 project managers. Some provided data for two projects; therefore, the total number of projects available for study was 1,539. After removal of outliers and bad data, the usable total was 1,386 projects. Outlier removal included the removal of projects which were cancelled or not completed (Table 13.2).

Responses were received from 60 countries. The largest percentage of respondents were from the USA (36%), followed by Canada and India. More than ten responses each were received from Australia, Spain, Brazil, Singapore, and Germany.

RESULTS AND ANALYSIS

To facilitate analysis, success questions were grouped into two overall measures of success. These were the measures of project success used throughout the analysis. They are:

Project Efficiency Measure

1. Meeting timeline goals
2. Meeting budget goals
3. Meeting scope and requirements goals

As project success is best judged by the stakeholders, especially the primary sponsor, we will use the following project success measure.

Project Success Measure

1. Sponsor assessment
2. Project team assessment
3. Client assessment
4. End user assessment

Finally, we had one question where the respondent, usually a project manager, gave an overall success rating to the project.

Overall Success Measure

1. Overall assessment of success by respondent

We completed a factor analysis that confirmed these groupings with the exception of the scope question. This question could fit with either the efficiency or success measure. This is in keeping with Shenhar et al. (1997) who stated that scope was the most important of the triple constraint for overall success. Since the result for scope was somewhat higher for factor efficiency, we will continue to use it as part of the efficiency measure.

Project Efficiency versus Project Success

The efficiency measure had a correlation of 0.60 with the project success measure and 0.58 with the respondents' self-reported overall success rating. The overall success rating was a single question asked of respondents: How successful overall was your project? Project success measure combined a number of questions and could be thought of as more accurate. It has a 0.87 correlation with the project success measure (Table 13.3).

Table 13.4 shows the correlation of time, cost and scope (the individual measures of project efficiency) with the measures of project success. The correlation with the overall project success rating and the project success measure is between 0.4 and 0.6. The highest correlation to overall success is with meeting scope goals, as we would expect (Shenhar et al., 1997). Scope is the most important part of efficiency for overall project success.

TABLE 13.3

Correlations between Project Success Measures

	Overall Project Success Rating	*Efficiency Measure*
Efficiency measure	0.58	
Project success measure	0.87	0.60

TABLE 13.4

Correlation of Individual Efficiency Measures to Project Success Measures

	Overall Project Success Rating	*Efficiency Measure*	*Project Success Measure*
Project time goals	0.51	0.88	0.51
Project budget goals	0.41	0.83	0.42
Scope and requirements goals	0.54	0.77	0.58

We can also look at how this relationship varies by industry (Table 13.5). It is interesting to note that efficiency is most highly correlated to project success in utilities, health care, and professional services. It is least correlated for government and high technology. This result for high technology may surprise some people though the result for government might not surprise many.

We can compare which components of efficiency were most important by industry (Table 13.6). Table 13.6 suggests budget goals and project success were most correlated for utilities, financial services, and health care. They were least important for government and retail. The finding for government is also in agreement with Kerzner (2009). Time goals were most correlated for construction and health care and least

TABLE 13.5

Correlation of Efficiency vs. Other Success Measures by Industry

	Overall Project Success Rating	*Project Success Measure*	*Valid N*
Construction	0.53	0.64	41
Financial services	0.64	0.68	257
Utilities	0.74	0.71	42
Government	0.47	0.41	152
Education	0.59	0.63	42
Other	0.51	0.58	157
High technology	0.50	0.52	223
Telecommunications	0.66	0.65	133
Manufacturing	0.69	0.69	122
Health care	0.61	0.69	113
Professional services	0.66	0.67	69
Retail	0.60	0.62	35

- All results above were significant at $p < .001$.

TABLE 13.6

Project Success Measures vs. Efficiency components by Industry

	Budget Goals	*Time Goals*	*Scope Goals*	*Valid N*
Construction	0.465	0.714	0.442	41
Financial services	0.496	0.566	0.660	257
Utilities	0.552	0.552	0.697	42
Government	0.304	0.300	0.424	152
Education	0.094*	0.572	0.701	42
Other	0.405	0.473	0.568	157
High technology	0.387	0.431	0.461	223
Telecommunications	0.396	0.557	0.684	133
Manufacturing	0.492	0.673	0.563	122
Health care	0.463	0.566	0.607	113
Professional services	0.450	0.564	0.662	69
Retail	0.349	0.474	0.702	35

Notes

* - Marked result was not significant at $p < 0.05$. All others were significant.

correlated with government and high technology. Scope goals were most correlated for education and utilities and least correlated with government and construction.

CONCLUSIONS

As suggested by many authors, overall project success is a much wider concept than the traditional so called iron triangle of project efficiency, (Atkinson, 1999). In this chapter, we have investigated to what extent project efficiency is correlated with overall project success. Through a survey of 1,386 projects we found that project efficiency is 60% correlated with project success. This falls to 51% if efficiency is defined as time and budget only. This supports the assertion that project efficiency is an important contributor to project success, but shows quite clearly that other factors contribute significantly as well. We can postulate that these other factors might include:

- Performance of the project's output post implementation, and achievement of the project's output and impact (Turner et al., 2010).

- Whether the project's output was what the stakeholders were actually expecting, or whether there was an omission in or misinterpretation of the specification.
- Risks that were not considered or changes to environment that were not anticipated (Munns and Bjeirmi, 1996; Thomas et al., 2008; Collyer and Warren, 2009).
- Acts of God beyond the project team's control.

Practical Implications

Whether we will ever be able to wean project practitioners off their beloved iron triangle, we cannot know. But this supports the work of Turner and Zolin (2012) that project managers need project control parameters that look beyond completing the scope of the project on time and within budget. Of the three efficiency components, scope is the most important while budget appears to be the least. Practitioners should be aware that when they plan and control the project that the broader success measures need to be taken into account and made part of the planning and control process. This will improve project and project manager perceived success especially over the long term. These results also demonstrate that practitioners cannot ignore project efficiency goals if they want to maximize overall success.

NOTE

1 This chapter is adapted from: Serrador, P., & Turner, J. R. (2015). The Relationship between project success and project efficiency. *Project Management Journal, 46*(1).

REFERENCES

Atkinson, R. (1999). Project management: cost, time and quality, two best guesses and a phenomenon, it's time to accept other success criteria. *International Journal of Project Management,* 17, 337–342.

Collyer, S. and Warren, C.M. (2009). Project management approaches for dynamic environments. *International Journal of Project Management,* 27, 355–364.

Cooke-Davies, T.J. (2002). The real success factors in projects. *International Journal of Project Management,* 20, 185–190.

Dvir, D., Raz T. and Shenhar, A. (2003). An empirical analysis of the relationship between project planning and project success. *International Journal of Project Management,* 21, 89–95.

Kerzner, H. (2009). *Project Management: A Systems Approach to Planning, Scheduling, and Controlling*, Wiley.

Munns, A. and Bjeirmi, B. (1996). The role of project management in achieving project success. *International Journal of Project Management*, 14, 81–87.

Prabhakar, G. (2008). What is project success: a literature review. *International Journal of Business and Management*, 26, 3–10.

Shenhar, A.J., Levy, O. and Dvir, D. (1997). Mapping the dimensions of project success. *Project Management Journal*, 28, 5–9.

Shenhar, A. and Dvir, D. (2007). *Reinventing Project Management: The Diamond Approach to Successful Growth and Innovation*, Harvard Business Press.

Thomas, M., Jacques, P.H., Adams, J.R. and Kihneman-Woote, J. (2008). Developing an effective project: planning and team building combined. *Project Management Journal*, 39, 105–113.

Turner, J.R. (2014). *The Handbook of Project-based Management*, 4th edition, New York: McGraw-Hill.

Turner, J.R., Huemann, M., Anbari, F.T. and Bredillet, C.N. (2010). *Perspectives on Projects*. London and New York: Routledge.

Turner, R. and Zolin, R. (2012). Forecasting success on large projects: developing reliable scales to predict multiple perspectives by multiple stakeholders over multiple time frames. *Project Management Journal*, 43, 87–99.

14

You've Just Inherited Someone Else's Project[1]: Now What?

Kate Davis[1] and Jeffrey K. Pinto[2]
[1]Cranfield School of Management, Cranfield, United Kingdom
[2]Black School of Business, Penn State, Erie, Pennsylvania, United States

INTRODUCTION

The majority of projects – even ultimately successful ones – run into significant problems during their development. While organizations have a variety of mechanisms at their disposal to correct projects that are experiencing difficulties, one of the most radical is replacing the project manager. Replacing a project manager "mid-stream" involves a major change to an ongoing project with the potential benefits of onboarding an individual with a different perspective or set of managerial and/or technical skills. When a project manager is replaced, it often sends the message that it is necessary to correct practice on troubled projects. It is also used to communicate to decision-makers and to team members the need for change and that processes and trust in governance needs to be re-established. This chapter discusses the decisions involved when replacing a project manager and the critical steps that the new project manager frequently undertakes in order to take control, assuage key stakeholders, and begin a series of remedial steps designed to bring the project back on track.

PROJECT MANAGER REPLACEMENT

In August 2018, London's £17.6 billion Crossrail Project announced the replacement of its former head, Simon Wright, after it was determined that the central section of the project (the Elizabeth line), scheduled to complete that December, would take at least another year before being ready for use (it was officially opened in May 2022). As part of the agreement by which the British government agreed to furnish an additional £650 m of funding, Wright was replaced by Mark Wild, the Managing Director of London underground, who remained in charge until the opening of the central section of Crossrail. As the most visible representative (and symbol) of the Crossrail project's difficulties, Wright was a symbol of the government's commitment to both complete the project and demand accountability for its delays, which ended up stretching out over two years past the original deadline.

The decision to replace a project manager during the execution phase of a project is one not taken lightly, nor is it likely to have insignificant consequences on the future viability of the project. Nevertheless, although such changes are surprisingly common (Dubber, 2015), we know surprisingly little about the reasons for project manager replacement and their potential impact on projects. Even when it is considered necessary to replace a project manager, it is often the case that it is done in a scattershot manner, with little forethought as to how the new project manager is expected to lead in this messy aftermath, both for the rest of the project team and the project itself. We suggest that it is possible to understand replacement mechanisms as a multilevel decision process, (1) identifying a set of antecedent "triggers" for replacement, (2) determining the effects of the actions that new project managers often undertake to promote their legitimacy and begin to "right the ship," and (3) final consequences, in the form of the impact their actions have on revitalizing the project and tracing a process for recovery. To understand this dynamic process, we intend to offer some practical advice for newly appointed project managers, tasked with bringing order out of the chaos of a mid-course project manager replacement. Our goal is to make clear that while the replacement decision can be risky and bring several new challenges for the incoming project manager, there are some useful remedial actions they can take to turn a potential disaster into a success story.

THE DECISION TO REPLACE THE PREVIOUS PROJECT MANAGER – WHY YOU HAVE BEEN SELECTED

To best understand answers to both the "why" question (why did we replace the original project manager?), as well as the "how" question

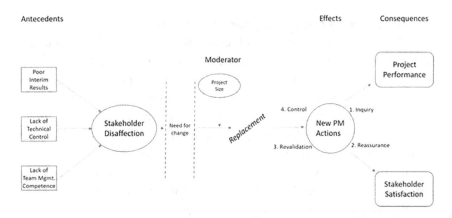

Antecedents   Effects   Consequences

FIGURE 14.1
Project manager replacement process model.

(how can it best be accomplished to minimize the disruption to the project and the rest of the team?), we studied the process at length, through interviews with a number of professionals within a diverse set of project-based industries – everything from defense to construction, new product development to service initiatives. The model we uncovered consisted of a series of steps that project organizations go through to identify the necessity of replacing the original project manager, and, once that decision has been made, the critical remedial activities undertaken by new project managers to minimize disruption, steady the project, and generate a renewed push to the finish line (Figure 14.1). For those who have experienced this replacement process and inherited a troubled project, you should find several of these steps familiar.

WHY THE ORIGINAL PROJECT MANAGER MIGHT HAVE BEEN REPLACED

There is a common misconception that all project manager replacement decisions arise from a similar root: the inability of the original project manager to successfully shepherd their project across the finish line. In other words, the first project manager failed at the job, put the project in jeopardy, and senior management was required to seek a maximal solution (i.e., firing this individual) as the only possible way to correct the situation. While it is certainly true that the decision to replace a

project manager is often made based on unsatisfactory project performance (in terms of schedule and cost) and/or the project manager's
personal characteristics (such as leadership style, technical expertise, and
communication skills), these are by no means the only reasons why it
makes sense to replace the original project manager. It's also important
to note that this replacement decision is often a complex one, involving
input from multiple stakeholders and decision-makers, and it is not
always clear-cut. The decision may also be influenced by organizational
factors, such as company culture, politics, and the perceived risks and
benefits of replacement. Let us look at some of the most common reasons
why the original project manager is replaced.

1. *For Midcourse Project Correction*

   Involuntary project manager replacement is a common corrective
   action during ongoing troubled projects. The decision to replace
   the project manager is to prevent ultimate project failure due to the
   chronic inability of meeting basic project targets of cost, time, or
   benefits realization. Poor project performance, and the consequent
   dissatisfaction of key stakeholders, are also main causes of project
   manager replacement. Finally, it is not uncommon for troublesome
   stakeholders to scapegoat the project manager as a convenient "fall
   guy" when the project is going through a rough patch. Under these
   circumstances, with a project reporting a steady litany of bad
   results, cost and schedule overruns, or technical errors, a convenient solution that sends a loud message to the parent organization
   and key stakeholders is to simply fire the project manager.

   Some of the common reasons for involuntary replacement have
   been associated with both the "hard" (technical) and "soft"
   (interpersonal) skills of the project manager. Technical issues
   include the inability to manage workload, project work not being
   up to standard (deficient quality standards), lack of technical skills,
   or the need for different skills for work packages. Whereas,
   interpersonal skills include the lack of relational (interpersonal)
   capabilities and leadership, relationship barriers and breakdowns,
   lack of personal and project team motivation, and poor performance as perceived by the client. Client disappointment is seen as the
   main trigger of the replacement process, where a perceived lack of
   competence from the current project manager is often flagged.
   Further, project indicators (e.g., time, cost, quality) will offer clear

evidential markers heralding an inevitable replacement point. Ultimately, the action to replace a project manager is often perceived as performance-related and normally does not represent a great shock among the internal stakeholders because evidence of poor performance might have been in the pipeline for some time.

2. *The Project Reaches Key Milestones That Require New Supervision*
   Although research suggests that the act of replacing the project manager is commonly dictated by poor project performance and key stakeholders' dissatisfaction (Dubber, 2015), it is also the case that this decision can be linked to the strategic direction of the project-based organization. Projects are social systems, and organization requirements and specifications might differ and change at each phase of the project-life cycle or at various key decision gates. New stakeholders, shifting political interests, and key actors in the supply chain can come into play at different points during project development. Therefore, at later, specific points in time, to reflect the needs of new social interactions, a new project manager may be judged to be better than their predecessor in managing, monitoring, and controlling the context in which these interactions are embedded.

   Consider, for example, a project that goes through several clear stages, including technical challenges, stakeholder and supplier coordination, political alliance-building, and final development. We found that for many firms, it is simply impossible to expect one project manager to be adept at this wide variety of duties. Consequently, they will routinely identify key milestones or project transition gates, at which time new project managers will assume responsibility for critical activities. In this situation, a technically-trained engineer may handle initial project management during the conceptual, proof-of-concept phase, and then hand the project over to a supply chain professional or business development specialist for later requirements. In this way, the changeover is planned, all parties are consulted throughout the project life cycle, and transitions are relatively seamless.

3. *Replacement as a Message for Change*
   The decision to opt for either an internal or external candidate to replace the project manager represents a serious issue for project decision makers. It is commonly believed that something "has" to be "done" when performance does not comply with the required

standards and expectations, and, just as is often the case with managers of failing sports teams, the project manager is seen as the first person to pay for this lack of performance.

We commented earlier that a perceived lack of critical skills can be the impetus to trigger the project manager replacement mechanism. Although the decision to source the new project manager internally or externally is very much context dependent – based on participants' experience, feelings, and reflections – a common understanding behind project manager replacement is that it provides a strong symbolic message for change to project stakeholders and the external world. In our research, we found that an important question to answer is: where will the new replacement project manager come from, inside or outside of the company? The implications of this question are very important because the common perception is that the replacement project manager is likely to be internal to the organization for *transitional* changes or, external to the organization for *transformational* changes.

If the goal of top management is to minimize disruption through a smooth transitional change aimed at taking corrective action to bring the project back on track, it is common to find the new project manager within the organization. This decision represents the most time and cost-effective solution, and is less risky, as most organizations often have a pool of skilled project managers already familiar with the environment within which the project operates. Consequently, sourcing internally usually will speed up the recovery process, by replacing the project manager with someone already familiar with the project management systems and processes and embedded in the organization culture. This decision is commonly viewed as less traumatic and safer for all other project team members; that is, new external project managers are seen as a threat to the project team through the potential for more widespread and disruptive shake-ups. On the other hand, the new, internal project manager is often judged to be the best option to overcome and mitigate relationship barriers and breakdowns. Both the project team and top management are more likely to collaborate with a familiar face from the internal existing organization.

There are also cases where the replacing project manager is external. Sometimes project requirements and/or its social dynamics did not match with the existing skills available within the organization; therefore, the replacement project manager had to be contracted externally. However, sourcing externally has been shown to be associated with the desire to bring major transformational change; that is, to have an unbiased perspective aimed at disconnecting with the way the project was managed by their predecessors.

A final goal is to send a strong message to the client and stakeholders to change the way people work, re-build credibility and motivation around a project deemed to be failing. In both cases, either the transitional or the transformational process has to be accompanied by a planned and well-organized handover.

4. *Re-establishing Processes and Trust through Handover – How the Project Size Matters*

Being part of a handover is complicated and requires careful planning and management in order to be as minimally disruptive as possible. As the new project manager, your aim is to ensure business continuity while forming a constructive environment and rebuilding trust among members of the project team. However, evidence suggests that there is no common agreement on how the handover process has been (or is actually) undertaken and managed in projects (Dubber, 2015; Vartiainen and Pirhonen, 2007). In fact, experience shows a mix of negative and positive handover events depending on a variety of factors, including the type and stage of the project, organization culture and the firm's appetite for change, sponsor pressure, and the urgency of the replacement itself. There are, however, some common beliefs on how the handover should be handled:

1. It should be well planned, following a structured process to assure a smooth project management transition;
2. As the new project manager, you need the organization to publicly support you in your new job, without underestimating the value that the old project manager brought in this transition. Depending on how the handover is presented, it can appear either chaotic and "ham-fisted" or carefully considered – even orchestrated;
3. Senior and executive management has to support you in taking the project through a brief step back, including allowing a collaborative overlap between you and the old project manager, and entering a short "reset" period; and

4. As the new project manager, you need the support of top management as you move through a careful onboarding and transition process to acknowledge a clear picture of the current situation of the project, understand the team's perspective, and repair both morale and stakeholder relationships. This process requires time, open and honest communication, and the willing collaboration of the old project manager over a short (but fundamental) period.

If possible, the goal is to make the replacement process as cooperative as possible, because your public partnering with the former project manager plays a crucial role in stakeholder acceptance. As you are attempting to build rapport, the project team is forced to recognize and adapt to a different way of working under new leadership and management styles. It is understandable that replaced project managers can get defensive about their legacy, leading to a tacit or even overt resistance to the transition, often accompanied by enlisting support from other team members. The lack of collaboration from these key players might, consequently, result in the loss of documentation and relevant (transparent) information vital for your goal of affecting positive change for the project. Handover involves a sometimes-steep learning curve. Successfully navigating this learning curve is on one hand very dependent on the support you receive from top management and, on the other hand, the trust and collaboration determined by the professionalism of the replaced project manager. Our research subjects agreed that if these conditions are in place, a smooth transition will support a corrective course of actions from the new project manager.

HOW TO BRING THE PROJECT BACK ON TRACK

Generally, the process of replacing a project manager can vary depending on the root cause of the problem and the specific circumstances.

However, some common steps that may be involved in the process include:

Step One: Identify the root cause of the problem: Before taking any action, it is important to understand why the project is failing. This may involve conducting a project audit or review to identify any problems or issues that are impacting the project's performance.

Step Two: Develop a recovery plan: Once the root cause of the problem has been identified, a recovery plan can be developed to address the issues and get the project back on track. This plan should include specific actions and milestones for achieving project goals and recovery.

Step Three: Assess the project team: The project team plays a critical role in the success of the project, so it may be necessary to assess their skills, experience, and workload. Identifying any gaps or issues with the team can help to address them and improve the project's performance.

Step Four: Communicate with stakeholders: Effective communication is essential for managing a failing project. Stakeholders should be kept informed of the project's status, progress, and any changes that are being made.

Step Five: Monitor and control progress: The progress of the project should be monitored and controlled to ensure that the recovery plan is being executed as intended and that the project is on track to meet its goals.

When commencing your new role, we suggest some steps to make the transition as smooth as possible. These steps came from our original study, as we found that successful new project managers (the replacements) often engaged in some common behaviors to make the transition as constructive as possible.

Step One: Inquiry – Your first job is to launch as thorough a review of the project as possible, including all critical performance metrics, interview data with stakeholders, technical reviews, team analysis, and so forth. In a nutshell, it is critical that you come into the project as the replacement project manager with full information: what is working, what is not working, potential trouble areas, etc. Getting a clear-eyed picture of the project also involves informational interviews with key team members, gaining their perspectives as a necessary first step to gaining their trust. Asking sufficient questions to give yourself the clearest possible picture of the project's status is a critical first step.

Step Two: Reassurance – When a project manager has been replaced, key stakeholders get very nervous. The natural questions to be asked include: how will these changes affect me and my assignment, will my position remain the same under the new boss, or even, will I like this person? Customers need reassurance that the project will be put back on track as quickly and painlessly as possible. Senior managers need to be kept in the loop because they are often the people responsible for providing the resources you need to finish the project. Our research found that effective transitioning project managers work hard to identify the key levers within their organization and reassure them as to the commitment to get things back in good shape as quickly as possible.

Step Three: Revalidation – Revalidation is the most challenging stage of the transition process, as bridges with the old project manager are now broken, and it is expected that you, the new project manager will begin to take corrective action. Remember that up to this point, replacement dynamics were aimed at smoothing the transition and alleviating the fears of key stakeholders; however, it is during revalidation that new goals or project team expectations are being clarified and implemented. Thus, you should now expect to see clashes occur as you start identifying new directions are given to the projects. The "people side" of the temporary organization might be affected as resources will be reallocated in order to rework the project. The project scope might need to be redefined based on the current needs of the organization, and a strong project governance system also has to be re-established. Team members will complain that "We did it other ways before," as they voice their nervousness over your new approach. Expect the conflicts that will arise during revalidation and make sure your message is clear and consistent through this part of the transition.

Step Four: Control – The four-step process concludes with a stage where your activities are focused on improving and refining project performance through controlling actions. It is necessary that you demonstrate value to the client and team by implementing and consolidating changes. In our study, subjects noted that a lack of control from the previous project manager was a common reason to explain deviations from the original plan, and it is incumbent on you to a better control process, including messaging, controlling project documentation and communication flow among key stakeholders. Extra meetings are often requested at this stage to fully communicate and lock-in the changes you are making going forward.

CONCLUDING THOUGHTS

The decision to take on the role of a replacement project manager during the execution phase of a project is one that should never be taken lightly.

The combination of administrative, interpersonal, technical, and organizational factors subject to upheaval during such a replacement explain why many organizations are hesitant to make this decision, opting instead for costly rework cycles after the fact. Further, the theories of escalation of commitment (Staw, 1981) and sunk costs (Garland, 1990) argue that choosing whether or not to take the major step of replacing a project manager remains one clearly resting in two decision arenas: technical project considerations as well as behavioral theory. Developing a clearer understanding of the process dynamics and well as the benefits and drawbacks of project manager replacement can aid organizations in making more clear-eyed decisions as they weigh present pain against future advantages. Most importantly, for those who are tasked as the new project manager with making this replacement process work, it is critical that you have the best information and action steps necessary to succeed.

NOTE

1 Portions of this chapter were adapted from: Davis, K., Di Maddaloni, F., and Pinto, J.K. (2023), "Drawing new cards or standing pat: Antecedents, dynamics, and consequences of project manager replacement," *IEEE Transactions on Engineering Management*, DOI: 10.1109/TEM.2021.3064609, (in press).

REFERENCES

Dubber, R.J., (2015), *Investigating the effects of replacing the project manager during project execution*, University of Johannesburg: Unpublished Master's Thesis.

Garland, H., (1990), Throwing good money after bad: The effect of sunk costs on the decision to escalate commitment to an ongoing project, *Journal of Applied Psychology*, 75, 728–731.

Staw, B.W., (1981), The escalation of commitment to a course of action, *Academy of Management Review*, 6, 577–587.

Vartiainen, T. and Pirhonen, M., (2007), 'How is Project Success Affected by Replacing the Project manager?', In: W. Wojtkowski, W.G. Wojtkowski, J. Zupancic, G. Magyar and G. Knapp (eds.), *Advances in Information Systems Development*, Boston, MA: Springer, 397–407.

15

Local Communities Stakeholder Defined: Identification and Categorization in Major Infrastructure Projects

Francesco Di Maddaloni
The Bartlett School of Sustainable Construction, University College, London, United Kingdom

BACKGROUND

To date, the understanding of megaproject impact on the local community level and how this can be minimized through a more inclusive approach to stakeholders' engagement remains marginal. Although understanding and minimizing the effect of major infrastructure developments on people and places can help to manage project benefits by moving towards more "community-inclusive" megaprojects (Bornstein, 2010); literature provides only a generic classification on the types of environmental (Melchert, 2007) and social (Vanclay, 2002) impact of infrastructure projects on communities.

On the other hand, the impact that the local community can exert on project results it is not new (Nguyen et al., 2019; Teo & Loosemore, 2017; van den Ende & van Marrewijk, 2019). MPIC projects seldom aligns project objectives with those of the local community (Derakhshan, 2020) and historically megaprojects have faced unpopularity and local opposition, with secondary groups trying to influence the implementation of facility development projects (Teo & Loosemore, 2014). This attitude is commonly labelled "Not in My Backyard" (NIMBY) syndrome, which is defined by Dear (1992, p. 288) as "the protectionist attitude of and oppositional tactics adopted by community groups facing an unwelcomed

DOI: 10.1201/9781003502654-15

development in their neighborhood," and it should be recognized as an expression of people needs and fears (Olander & Landin, 2008).

While stakeholder theory recognizes the growing importance of communities as a new class of stakeholders, the issues of their identification and prioritization has never been fully resolved (Crane & Ruebottom, 2011; Di Maddaloni & Davis, 2018). The literature review revealed conflicting definitions and conceptualizations of the local community. Although Webber (1963) first set the stage, from a perspective of a construction project, for broadening the notion of community away from purely place-based definitions, community refers to a multitude of overlapping, competing and conflicting interests groups, which shift over the project life cycle and whose interests are potentially affected by that project (Teo & Loosemore, 2011). The local community cannot be treated as a single homogeneous, easily identified group (Atkinson & Cope, 1997; Skerratt & Steiner, 2013), and in the stakeholder management literature, the concept of community has been left constantly unclear and undefined.

Due to the physical impact of megaprojects, this study emphasizes the traditional view based on geography, or place-based communities which, centered on the physical proximity of the members to project developments (Dooms et al., 2013; Driscoll & Starik, 2004). It is believed that managing the local community will help to manage benefits (Davis et al., 2021; Di Maddaloni & Davis, 2017), by aligning MPIC objectives and interests with those of the local community and enhancing a shared view of project objectives to aid in achieving better project performance in terms of value creation and distribution (Di Maddaloni & Sabini, 2022; Gil & Fu, 2021). However, it is highlighted that if there is no clear definition, it is not possible to determine whether the relevant components of the community of place have been correctly identified and, consequently, whether a stakeholder analysis has been successfully accomplished.

By exploring the literature, it is evident how stakeholder engagement practices at the local level of MPIC projects still are not fully captured by practitioners and academics alike in the stakeholder management arena. Therefore, the key contribution of this chapter is to provide an answer to the following research question:

How the local community stakeholder is perceived, identified, and categorized by project managers in MPIC projects?

DATA AND FINDINGS

Fifteen of the most representative MPIC projects in UK were discussed and contextualized, resulting in a total of 19 interviews with key people in the construction industry, consisting of nine communication managers, six project managers and four senior managers. The analysis of the 19 interviews produced more than 900 initial codes. When analyzing the interviews, it became evident that the interviewees' feelings, perceptions, and understanding of the topic under scrutiny resulted in two sets of themes that captured the most important elements of the data regarding the project management perception of the stakeholder local community in MPIC projects: (1) MPIC, local communities, and stakeholder management – a negative bond and (2) defining the identity of the local communities.

MPIC, Local Communities, and Stakeholder Management: A Negative Bond

The management of disempowered stakeholders (which do not have formal contractual relationship with, or direct legal authority over, the organization) in MPIC projects starts and operates in a negative bond. Based on the perceptions of the interviewees, the findings indicate that the MPIC impact at the local level is perceived negatively by project managers of megaprojects. In fact, general beliefs from the interviews consider the negative consequences of MPIC to local communities exceeding the positive aspects of these developments. This is mainly associated with the disruption that these projects typically have in peoples' day-to-day lives. Some of the common negatives that emerged from the interviewees were noise, dust, pollution, lighting, traffic congestion, land acquisition, changes in landscape, and unaffordable rent due to increased value of the property.

Although it is recognized that the local community cannot be treated as a single homogeneous group, but it has to be seen as multiple separate components with their own needs, expectations and attitudes towards the entire project life cycle; it is perceived that MPIC projects are not generally welcomed by local communities' groups. In fact, according to participants' feelings, too often local communities see MPIC projects as a threat rather than an opportunity. Project managers sense that local communities have a general disbelief towards MPIC projects which can cause them shock, fear, and affliction. In the same way, participants' belief is that usually the local communities start the engagement process highly skeptical and with a negative mindset on how MPIC projects are ultimately going to have a negative effect on them. What is perceived is that there are pre-existing prejudices behind an organization's strategy and people suspect that engagement is all about manipulation. Stakeholder engagement and consultation at the local level is often perceived as paid lips service, where decisions are already made and cannot be influenced or changed in any way.

The negative dynamics, in which secondary stakeholder management operates at the local level of MPIC projects, is also reflected in the project management perception of the local communities' stakeholder. Although the interviews indicate that the recognized benefits are different when engaging with this class of stakeholders, such as increasing efficiency by having a smoother ride towards project completion, saving time, money, increasing organization's reputability, gain local intelligence/knowledge and increase the benefits of MPIC projects; participants perceive the stakeholder local communities negatively. Organizations often do not want their project managers to deal with the external world and they are primarily looking at the local communities as a risk from a perspective of the implications of delays around public consultation. Among others, the local communities were also defined as an irritant, narrow minded, without vision, and driven by self-interests (Figure 15.1).

Defining the Local Community Identity

Different cohorts make up the local community as the norm rather than exception in MPIC projects. These cohorts are the residents community,

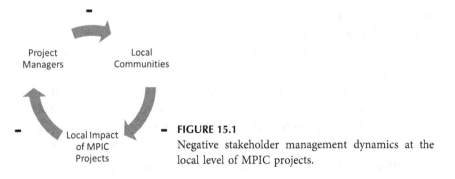

FIGURE 15.1
Negative stakeholder management dynamics at the local level of MPIC projects.

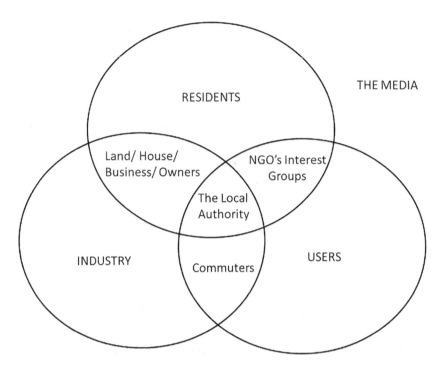

FIGURE 15.2
Identification of the local communities' stakeholder in MPIC projects.

the businesses community and the users community which, as shown in Figure 15.2, will form the shape for other sub-cohorts namely the property owners (i.e., land, house, business), the customers (i.e., commuters, road users) and the NGO's interest groups (i.e., schools, churches, local association groups). The media, which sits outside the local community cohorts but has a direct influence to them, has to be considered by managers. Although each of these groups can show different interests, levels of engagement and opposition or support at each phase of the project life cycle, to which managers are asked to respond, the results from the interviews show how the local authority is considered the most influential cohort of the component local communities.

Although the physical identification of the local communities stake-holders in MPIC projects does not represent a simple exercise as its conceptualization is really dependent on the nature and type of project, its geographical location (in the city or out of the city) and peoples demographic backgrounds such as culture/race, gender, age, welfare;

common behavioral attitudes, and actions of the local communities in regards to MPIC projects developments have emerged from the interviews which allow their categorization.

Six distinct categories of local communities relevant to stakeholder theory were identified. By putting emphasis on geography (proximity), interest (opinions), perceived impact and benefits of the MPIC project, these include "community of interests," the "silent majority," the "opportunists," the "negatively affected," the "beneficiary" and the "unconditional opponents."

1. The *"community of interest"* refers to individuals that are unified by a common purpose or interest and may not be in close proximity to the organization's operation. These can be commuters or road users, which have little or no emotional attachment and are not sensitive. This category might overlap with the 'silent majority' due to either their proximity or opinions.

2. The *"silent majority"* refers to individuals that, although they are in the proximity of project development, they do not express direct or official opinions about the project and they do not engage with the organization. These can be people who are dis-engaged due to cultural barriers or having no time. They have a small to medium emotional attachment and are minimally sensitive.

3. The *"opportunists"* refer to individuals that might be in the proximity of the project and have direct opinions, but they have no direct impact (either positive or negative) on the project development. They are motivated by self-interest and exploiting opportunities to get something from the project, but they do not have an honest *bona fide* interest. These can be people with a medium to high emotional attachment and are sensitive. This category might overlap with the "negatively affected" due to their proximity and/or emotional attachment.

4. The "negatively affected" refers to individuals that receive no direct or tangible benefit from the project. They might be in the close proximity or have an opinion, but the impact of the project is perceived to outweigh the benefits. These can be landowners, house owners, small business who have no ability to transfer their operations somewhere else. They have a very high emotional attachment and are very sensitive. This group is very keen to oppose the project.

5. The *"beneficiary"* community refers to individuals that receive direct or tangible benefits from the project. They might be in close proximity, have an opinion, and the impact of the project is perceived to be positive. These can be residents, businesses and users driven by the long-term vision for a positive change. They have a very high emotional attachment and are very sensitive. This group is very keen to support the project.

6. The *"unconditional opponents"* refer to individuals that might or might not be in the proximity to the project, have an opinion or impact, but they do not want *a priori* the project to happen. They are highly oppositional and difficult to engage in constructive dialogue, whose purpose appears to be the short-term goal of disruption, rather than any problem resolution. They might or might not have an emotional attachment or be sensitive, but they can exert highly negative influence on the other groups.

Based on the above characteristics, the conceptual categorization of the local communities in MPIC projects is shown in Figure 15.3.

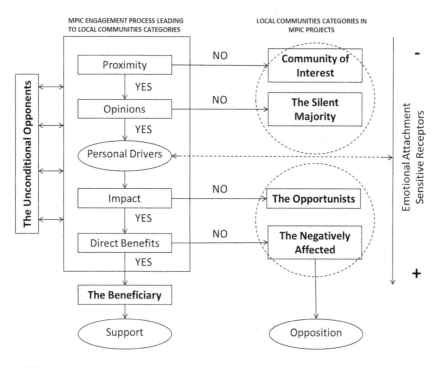

FIGURE 15.3
Categorization of the local communities' stakeholder in MPIC projects.

IMPLICATIONS FOR PROJECT MANAGERS

Stakeholder management at the local level of MPIC projects operates within negative dynamics, where local communities groups are perceived as a risk by project managers and, on the other hand, the engagement process applied to the local communities groups seems to be perceived as merely a paid lips service, where decisions are made and cannot be changed in any way (Di Maddaloni & Derakhshan, 2019). This, inevitably, requires project managers to give extra managerial effort in terms of time and resources, such as building relationships through a face-to-face approach, also claimed by Hart & Sharma (2004).

Although major important steps have been made in the last 20 years, MPIC projects often have a national agenda and their impact at the local level is often perceived negatively. Expanding the findings of Dooms et al. (2013), this work asserts that the impact, salience, and management strategies of disempowered stakeholders in MPIC projects are assessed by proximity, which also represents the most common conceptualization that project managers have of the local communities' stakeholders in MPIC projects. The interviews reinforced that the local communities' stakeholders in MPIC projects cannot be treated as a single homogenous group (Skerratt & Steiner, 2013), and their physical identification and assessment is dependent on the nature and type of MPIC project. However, three main cohorts of local communities in MPIC projects, as the norm rather than exception, have been conceptualized and can be used by project managers for a better identification of this class of stakeholders. These cohorts include the residents' community, the businesses community and the users' community (the media is positioned outside). These cohorts create the bases for other sub-groups to which the local authority is the representative. The local authority assumes a position of control to other groups, and it has been considered the most influential cohort which organizations aim to work closely with.

Evidence suggests the difficulties when identifying the local communities involved in MPIC projects. However, common themes emerged in their behavioral attitudes and actions towards MPIC projects which facilitated categorization. This study expands and integrates the work of Dunham et al. (2006) by identifying six categories of the local communities' stakeholder. These categories include "community of interests," the "silent majority," the "opportunists," the "negatively

TABLE 15.1

Recommended Stakeholder Engagement Strategies at the Local Level of MPIC Projects

		Community of Interest	The Silent Majority	The Opportunists	The Negatively Affected	The Beneficiary	The Unconditional Opponents
PROXIMITY							
Behavior and Actions	Proactive	X	X	X	X	X	X
	Oppositional		X	X	X	X	X
Consultation	Formal	X	X	X			X
	Informal (face-to-face)		X	X	X	X	
Engagement Aim	Collaboration		X			X	
	Cooperation	X	X	X	X		
	Containment				X		X

affected," the "beneficiary" and the " unconditional opponents." Drawing from local communities perceived behaviors and attitudes can help managers to allocate the right resources and effort on those stakeholders possessing a proactive, neutral or oppositional perception about the project. The aim is to maximize local communities' positive inputs towards cooperation and collaboration, or minimize their negative attitude by containment, as shown in Table 15.1.

In this chapter, it has been elucidated that the organization must take account of the effects of their behavior upon those who live in close physical *proximity* to their operations. Based on their proximity and perceived impact (positive or negative) of the MPIC project, different communities groups can show a *proactive* or *oppositional* behavior towards the project. It is the responsibility of the project manager to monitor and anticipate any shifting attitudes towards the entire project life cycle and develop appropriate strategies. According to Dunham et al. (2006) three distinct strategies can be adopted when approaching communities groups: *collaboration, cooperation,* and *containment.* While the aim of collaboration is to support stakeholder development through open and trust-based interaction and building shared vision for the project; cooperation strategy is more about building a win-to-win solution along a cordial and reciprocal interaction which will lead on sharing selective information through an ongoing dialogue. On the other hand, containment strategy has a process focused on identifying and monitoring which aims to minimize potential damages by stakeholder groups and where the nature of the interaction is often adversarial.

Although they do not track perfectly, these strategies have different objectives and can be broadly mapped against the six categories of community presented in this chapter. Of course, each project and stakeholder's personal drivers are distinct, and it remains the responsibility of the practitioner to develop a more specific understanding of each stakeholder group and then determine the appropriate strategy.

REFERENCES

Atkinson, R., Cope, S. (1997). Community participation in urban regeneration in Britain. In Hoggett, P. (Ed.), *Contested Communities.* Bristol Policy Press, pp. 201–221.
Bornstein, L. (2010). Mega-projects, city-building and community benefits. *City, Culture and Society,* 1, pp. 199–206.
Crane, A., Ruebottom, T. (2011). Stakeholder theory and social identity: Rethinking stakeholder identification. *Journal of Business Ethics,* 102, pp. 77–87.

Davis, K., Pinto, J.K., Di Maddaloni, F. (2021). Chapter II: Assessing the value and benefits of major infrastructure projects. *Routledge Handbook of Planning and Management of Global Strategic Infrastructure Projects*, 1st edition, Routledge, London and New York.

Dear, M. (1992). Understanding and overcoming the NIMBY syndrome. *Journal of the American Planning Association*, 58, 288–300.

Derakhshan, R. (2020). Building projects on the local communities' planet: Studying organizations' care-giving approaches. *Journal of Business Ethics*. 10.1007/s10551-020-04636-9

Di Maddaloni, F., Davis, K. (2017). The influence of local community stakeholders in megaprojects: Rethinking their inclusiveness to improve project performance. *International Journal of Project Management*, 35(8), 1537–1556

Di Maddaloni, F., Davis, K. (2018). Project manager's perception of the local communities' stakeholder in megaprojects. An empirical investigation in the UK. *International journal of Project Management*, 36(3), 542–565.

Di Maddaloni, F., Derakhshan, R. (2019). A leap from negative to positive bond. A step towards project sustainability. *Administrative Sciences*, 9(2), p. 41.

Di Maddaloni, F., Sabini, L. (2022). Very important, yet very neglected: Where do local communities stand when examining social sustainability in major construction projects?. *International Journal of Project Management*, 40(7), pp. 778–797.

Dooms, M., Verbeke, A., Haezendonck, E. (2013). Stakeholder management and path dependence in large-scale transport infrastructure development: The port of Antwep case (1960–2010). *Journal of Transport Geography*, 27, 14–25.

Driscoll, C., Starik, M. (2004). The primordial stakeholder: Advancing the conceptual consideration of stakeholder status for the natural environment. *Journal of Business Ethics*, 49, 55–73.

Dunham, R., Freeman, R.E., Liedtka, J. (2006). Enhancing stakeholder practice: A particularized exploration of community. *Business Ethics Quarterly*, 16(1), 23–42.

Gil, N., Fu, Y. (2021). Megaproject performance, value creation, and value distribution: An organizational governance perspective. *Academy of Management Discoveries*, 8(2), 224–251.

Hart, S.L., Sharma, S. (2004). Engaging fringe stakeholders for competitive imagination. *Academy of Management Executive*, 18(1), 7–18.

Melchert, L. (2007). The Dutch sustainable building policy: A model for developing countries? *Building and Environment*, 44(2), 893–901.

Nguyen, T.H.D., Chileshe, N., Rameezdeen, R., Wood, A. (2019). External stakeholder strategic actions in projects: A multi-case study. *International Journal of Project Management*, 37(1), 176–191.

Olander, S., Landin, A. (2008). A comparative study of factors affecting the external stakeholder management process. *Construction Management and Economics*, 26, 553–561.

Skerratt, S., Steiner, A. (2013). Working with communities of place: Complexities of empowerment. *Local Economy*, 28(3), 320–338.

Teo, M., Loosemore, M. (2011). Community-based protest against construction projects: A case study of movement continuity. *Construction Management and Economics*, 29(2), 131–144.

Teo, M., Loosemore, M. (2014). The role of core protest group members in sustaining protest against controversial construction and engineering projects. *Habitat International*, 44, 41–49.

Teo, M., Loosemore, M. (2017). Understanding community protest from a project management perspective: A relationship-based approach. *International Journal of Project Management*, 35(8), 1444–1458

van den Ende, L., van Marrewijk, A. (2019). Teargas, taboo and transformation: A neo-institutional study of community resistance and the struggle to legitimize subway projects in Amsterdam 1960–2018. *International Journal of Project Management*, 37(2), 331–346.

Vanclay, F. (2002). Conceptualizing social impacts. *Environmental Impact Assessment Review*, 22(3), 183–211.

Webber, M. (1963). Order in diversity: Community without propinquity. In Wingo, L. (Ed.), *Cities and Space*. Baltimore: John Hopkins Press.

16

What Planning Effort Optimizes Project Success?[1]

Pedro M. Serrador
Northeastern University, Toronto, Ontario, Canada

"By failing to prepare, you are preparing to fail."

– Benjamin Franklin

INTRODUCTION

In this chapter we investigate the impact of project planning and project plans on project success. Does better project planning lead to more successful outcomes on projects? Traditional project management is based to a large extent on conjecture, with little empirical evidence in support of some of the memes (Turner, Huemann, Anbari, & Bredillet, 2010). Project planning is one such meme. Received wisdom is that planning is very important and the more effort that is put into the planning process, the better the project plans and the more successful will be the project (Dvir, Raz, & Shenhar, 2003). Time spent on planning activities will reduce risk and improve success. On the other hand, inadequate planning will lead to a failed project (Thomas, Jacques, Adams, & Kihneman-Woote, 2008). If poor planning has led to failed projects, then perhaps trillions of dollars have been needlessly lost (Sessions, 2009). Our survey of the literature suggested there is a relationship between the amount of project planning and the quality of project plans, and between both of those and project success. But is there an optimum amount of planning and how much is too much? We believe this relationship needs to be clarified.

DOI: 10.1201/9781003502654-16

LITERATURE REVIEW

Project Success

Before we can discuss the impact of the project planning phase on success, we need to define what we mean by project success. Unfortunately, as Pinto and Slevin (Pinto & Slevin, 1988) note "There are few topics in the field of project management that are so frequently discussed and yet so rarely agreed upon as the notion of project success." (p 67). Shenhar and Dvir (2007) suggest five measures of project success:

1. project efficiency
2. impact on the team
3. impact on the customer
4. business success
5. preparing for the future

Thomas *et al.* state (2008), "Examples abound where the original objectives of the project are not met, but the client was highly satisfied," as well as the reverse (p106). Ultimately, whether or not the latter is achieved is a subjective judgment by key stakeholders (Serrador & Turner, The Relationship between project success and project efficiency, 2015). While the measure of project success has in the past focused on tangibles, current thinking is that ultimately project success can best be judged by the primary sponsor (Cooke-Davies, 2002) and will be based on how well they judge that the project meets the wider business and enterprise goals.

We endeavor to work with a more simplified model. In this chapter, we refer to:

- *project efficiency*: completing the desired scope of work on time and within budget while meeting scope goals
- *project success*: meeting wider business, strategic, and enterprise goals

Project Planning

Mintzberg describes planning as the effort to formalizing decision making activities through decomposition, articulation, and rationalization

(Mintzberg, 1994). For the purpose of this chapter, we will use these definitions:

- *planning phase:* the phases and associated effort that comes before execution in a project
- *planning effort:* the amount of effort in work hours expended in planning

Thomas et al. (2008) state, "the most effective team cannot overcome a poor project plan" (p. 105) and projects which started down the wrong path can lead to the most spectacular project failures. Morris (1998) similarly argues that "The decisions made at the early definition stages set the strategic framework: Get it wrong here, and the project will be wrong for a long time" (p. 5). Thus, there is a recurring theme that planning is inherently important to project success; one could argue that without it project management would not exist.

Pinto and Prescott (1988) found that a schedule or plan had a correlation of 0.47 with project success, while technical tasks had a correlation of 0.57 and mission definition a correlation of 0.70. Planning was found to have the greatest impact on the following success criteria: perceived value of the project ($R^2 =. 35$); and client satisfaction ($R^2 = .39$).

Dvir and Lechler (2004) found quality of planning had a +0.35 impact on R^2 for efficiency and a +0.39 impact on R^2 for customer satisfaction. Dvir et al (Dvir, Raz, & Shenhar, An empirical analysis of the relationship between project planning and project success, 2003) noted the correlation between aspects of the planning phase and project success. Zwikael and Globerson (2006) noted the following, "organizations, which scored the highest on project success, also obtained the highest score on quality of planning." (p. 694)

Reasons Not to Plan

Andersen (1996) questions the assumption that project planning is beneficial from a conceptual standpoint. He asks, "How can it be that project planners are able to make a detailed project plan, when either activities cannot be foreseen or they depend on the outcomes of earlier activities?" (p. 89). Bart (1993) makes the point that in research and development (R&D) projects, too much planning can limit creativity.

Collyer et al. (2010) describe examples of failed projects such as the Australian submarine and the Iridium satellite projects. They say, "While useful as a guide, excessive detail in the early stages of a project may be problematic and misleading in a dynamic environment" (p. 109). Collyer and Warren suggest that, in dynamic environments, creating detailed long-term plans can waste time and resources and lead to false expectations (Collyer & Warren, 2009). Aubry et al. (2008) note that for one project management office (PMOs) they studied, overly rigorous planning processes resulted in an impediment to rapidity. Flyvbjerg et al. (2002) highlight that senior management can choose not to use the estimates from the planning phase.

For this research, we undertook a global survey of projects. People from over 60 countries answered the survey. A total of 859 people provided data on 1,386 projects. The largest numbers came from the US, 313 (36.5%); India, 59 (6.9%); Canada, 57 (6.6%); and Australia, 19 (2.2%). Some 183 (21.3%) chose not to answer the question. Although there was a preponderance of responses from North America, there was good representation from the whole world.

Inductive analysis was used to find the relationship between planning and success. In general, the simplest relationships were tested first and then testing continued using progressively more involved techniques. The typical progression is to use correlation analysis to understand if there is a relationship followed by linear regression to see if there is a dependent relationship. This is followed by a non-linear regression if a significant linear relationship is not discovered. Finally, MHRA (Moderated Hierarchical Regression Analysis) was used to understand how this relationship is impacted by moderating variables (Sharma, Durand, & Gur-Arie, 1981).

RESULTS AND ANALYSIS

We next examined how the quality of planning impacted project success by completing a regression of success versus the planning quality factor. We found a statistically significant relationship with a low $p < 0.0001$ between the planning quality factor and overall success ($p < 0.05$ is the standard so this result is considered quite good). In addition, there is a strong R^2 of 0.265. This result is in broad agreement with the average R^2 reported in the literature.

We then wanted to analyze the effort spent planning rather than the quality of the planning done.

Planning effort vs. Success

To start the effort impact analysis, we examined the relationship between planning effort and project success rating. The planning effort is shown as the percentage of total effort spent on planning. The planning effort index is the amount out of 1.00. We found that in general the planning effort increases within the success category. The exception is the failure category that showed the highest mean planning effort of any group.

To start the analysis, we examined the relationship between planning effort and project success rating (Table 16.1). We can see that in general the planning index increases within the success category. The exception is the failure category that shows the highest mean planning effort index of any group.

The ANOVA analysis did not show a statistically significant relationship. By looking at these means it appeared that a simple linear relationship did not exist. These data were now plotted to get a visual picture of the relationship (Figure 16.1). Looking at this graph, we can see the lowest amount of effort was typically spent on projects deemed not fully successful. In this case, one can hypothesize that inadequate planning impacted project success. Projects deemed outright failures reported the mean highest percentages of upfront project planning.

TABLE 16.1

Planning Effort Index and Project Success Rating for All Projects

	Planning Effort	*Number of Projects*
Failure	16.6%	98
Not fully successful	14.3%	259
Mixed	15.2%	345
Successful	15.4%	451
Very successful	15.8%	233
All groups	15.3%	1386
p(F)	*0.178*	

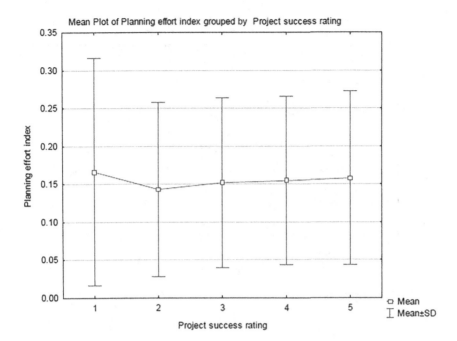

FIGURE 16.1
Mean plot of planning effort index by project success rating with error bars (where 5 is a highly successful project).

Based on Figure 16.1, it was decided to review the data with an assumption that the relationship between the effort index and project success is not linear but was a curve. An inverted-U curve also fits with some of the findings of the literature review (Figure 16.2).

One can see that there are a very wide range of projects in the sample with different planning characteristics and success levels. What is striking is when a computer analysis is done, an inverted U curve appears though no curve is apparent to the naked eye. These types of analyses are, of course, not possible without computers. This initial curve is rather shallow showing a small effect.

When we completed more advanced analysis (MHRA) which separated out the effects of confounding moderators, a more significant relationship between planning effort and project success was been uncovered, with $R^2 = 0.145$, $p < 0.001$ (Figure 16.3).

The point we must take away is that for the average project, doing no up-front planning will reduce its success rating by 0.5 to 1 success levels. That is, turn a successful project to an average project, or an average

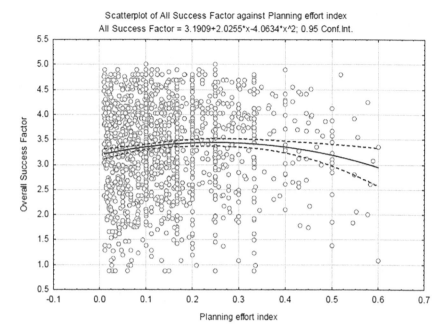

FIGURE 16.2
Scatterplot and curve fitting for overall success factor versus planning effort index.

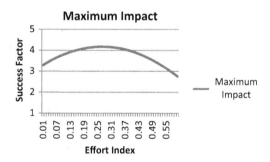

FIGURE 16.3
Planning effort index vs. success for maximum moderator values.

project to an unsuccessful one. The impact on projects that have an overly long planning phase is even more severe. One can note from this graph that the y intercept indicates that for zero upfront planning projects still show average success. However, these curves refer to averages of variable projects with numerous moderators in play. Many projects also do substantial planning during execution.

TABLE 16.2

Optimum Planning Effort Index Values by Success Measures

	Success Measure	*Stakeholder Success Measure*	*Efficiency Measure*	*Mean*	*Actual Reported Planning Index Average*
Planning effort index	25.5%	24.8%	25.0%	25.1%	15.3%

We already had a curve of planning effort versus success. Was there any useful guidance could we give to project managers, program managers or executives? If we look at the curve for the relationship between planning effort and success, I thought there should be a maximum at the peak. Since there is a quadratic relationship between planning effort and success measures, it was possible to calculate a maximum to the resulting quadratic curve. If you remember from high school, for a quadratic equation of the form $Ax^2 + Bx + C$, the formula for the maximum is B/2 A. After completing this calculation for the overall success measure, this value was found to be 0.249 or approximately 25% of effort spent on planning prior to execution. This is a very interesting result that indicates on average projects which spend approximately 25% of their effort on the planning phase, are the most successful.

We now calculate the planning effort index value which maximized the stakeholder success measure and efficiency measure in the same way we did for the success measure. Table 16.2 below summarizes those calculations.

These results are interesting from a number of viewpoints. They are in line with the approximately 20–33% effort spent on planning identified in the literature. The three results are also with 0.03 of each other, which helps to validate the research methodology. Finally, the 25% is higher than the average of 15% planning efforts reported by survey respondents.

We can see from this table that the optimum planning amounts are relatively consistent between the three success measures. In addition, optimum planning values were calculated on subsets of the data. The results were often within the .20–.25 range.

Subgroups were also analyzed to confirm optimum planning levels (Table 16.3). Below is a summary of some of those analyses where results differ from the average. The statistical significance level ($p <= .05$) was relaxed to $p \sim .10$ to allow a broader view. The results show some of the variation between industries. Retail, for example shows a very high

TABLE 16.3

Summary of Subgroups Analysis for Optimum Planning Levels

	P	Valid N	Average Planning Effort	Optimum Planning Level
Region – North America	0.031	756	0.151	0.226
Team type – International	0.03	442	0.149	0.216
Industry – Professional Services	0.087	54	0.139	0.251
Industry – Education	0.118	42	0.132	0.210
Industry – Government	0.105	152	0.126	0.147
Industry – Retail	0.108	30	0.173	0.509

planning optimum perhaps because for retail projects, the majority of work really is in the planning. There is little to build in execution compared to construction and IT, for example. Professional services plan less than average. This may be because, change requests can be a source of additional income in some cases. Interestingly, the optimum planning level for government projects was quite low. One can speculate that they are less dependent on up front planning and perhaps more dependent on other success drivers such as change management or stakeholder management. We argue this is in keeping with the work of Flyvbjerg et al. (2002).

SUMMARY

This research did confirm the relationship between planning effort and project success and that a quadratic relationship exists between the percentage of effort spent planning and project success. The planning effort has a stronger link with overall success than with project efficiency. This may indicate that shortening planning cycles impact projects by reducing their final value to the company and stakeholders even though managers may still be able to deliver them on time and budget.

Planning is important to project success as numerous authors have previously written. It is clear from this research that the average project is not spending enough time on upfront planning to maximize success.

This should not be surprising to researchers or practitioners; it appears that in industry not enough planning is being done and that if longer planning phases were the norm, there would be higher overall project success. The inverted-U shaped relationship between planning effort and success is significant and should be considered in future research. Also, what is not widely known is that projects that spend too much time planning have a much lower success on average than other project.

The planning phase effort does not impact all aspects of success equally. The planning phase effort has the strongest relationship with overall project success. Reducing the effort spent on the planning phase may impact projects by reducing their final value to customers, stakeholders, and the company. This may be the case even though managers may still be able to deliver them on time and within budget.

The phenomenon that projects may not be planning adequately could be a factor in the high project failure rates reported in the literature (Sessions, 2009; The Standish Group, 2011). It is recommended that projects consider doing more planning upfront both for traditional projects and for agile projects. However, projects with too long a planning phase were also found to have lower success ratings. Projects that schedule more than 25% effort on the upfront planning phase should be reviewed for progress and risk factors. Over-planning could be a symptom of a project that is too complex to deliver successfully, a lack of firm requirements or of a team that is not experienced enough in this project area: all of which could potentially lead to a failed project. Finally, 25% appears to be rule of thumb for the optimum amount of effort spent planning.

NOTE

1 This chapter is adapted from: Serrador, P., & Turner, J. R. (2015). What is enough planning? Results from a global quantitative study. *IEEE Transactions on Engineering Management, 62*(4), 462–474.

REFERENCES

Andersen, E. S. (1996). Warning: activity planning is hazardous to your project's health. *International Journal of Project Management, 2*(14), 89–94.

Aubry, M., Hobbs, B., & Thuillier, D. (2008). Organisational project management: an historical approach to the study of PMOs. *International Journal of Project Management, 26*(1), 38–43.

Bart, C. (1993). Controlling new product R&D projects. *R\&D Management, 23*(3), 187–198.

Collyer, S., & Warren, C. M. (2009). Project management approaches for dynamic environments. *International Journal of Project Management, 27*(4), 355–364.

Collyer, S., Warren, C., Hemsley, B., & Stevens, C. (2010). Aim, fire, aim – project planning styles in dynamic environments. *Project Management Journal, 41*(4), 108–121.

Cooke-Davies, T. J. (2002, #apr#). The real success factors in projects. *International Journal of Project Management, 20*(3), 185–190.

Dvir, D., & Lechler, T. (2004, #jan#). Plans are nothing, changing plans is everything: the impact of changes on project success. *Research Policy, 33*(1), 1–15.

Dvir, D., Raz, T., & Shenhar, A. (2003). An empirical analysis of the relationship between project planning and project success. *International Journal of Project Management, 21*(2), 89–95.

Flyvbjerg, B., Holm, M. S., & Buhl, S. (2002). Underestimating costs in public works projects: error or lie? *Journal of the American Planning Association, 68*(3), 279–295.

Mintzberg, H. (1994). *The rise and fall of strategic planning: reconceiving roles for planning, plans, planners.* Englewood Cliffs, NJ, USA: Prentice Hall.

Morris, P. W. (1998). Key issues in project management. In J. K. Pinto (Ed.), *Project {Management} {Institute} {Project} Management handbook.* Newtown Square, PA: Project Management Institute.

Pinto, J. K., & Prescott, J. E. (1988, #mar#). Variations in critical success factors over the stages in the project life cycle. *Journal of Management, 14*(1), 5–18.

Pinto, J. K., & Slevin, D. P. (1988). Project success: definitions and measurement techniques. *Project Management Journal, 19*(1), 67–72.

Serrador, P., & Turner, J. R. (2015). The relationship between project success and project efficiency. *Project Management Journal, 46*(1), 30–39.

Sessions, R. (2009). *The IT complexity crisis: danger and opportunity.* Tech. rep., ObjectWatch, Inc. Retrieved from http://www.objectwatch.com/whitepapers/ITComplexityWhitePaper.pdf

Sharma, S., Durand, R., & Gur-Arie, O. (1981). Identification and analysis of moderator variables. *Journal of Marketing Research,* 291–300.

Shenhar, A., & Dvir, D. (2007). *Reinventing project management: the diamond approach to successful growth and innovation.* Harvard Business Press.

The Standish Group. (2011). *CHAOS Manifesto 2011.* Accessed June, 2011, The Standish Group. Retrieved from http://standishgroup.com/newsroom/chaos_manifesto_2011.php

Thomas, M., Jacques, P. H., Adams, J. R., & Kihneman-Woote, J. (2008). Developing an effective project: planning and team building combined. *Project Management Journal, 39*(4), 105–113.

Turner, J. R., Huemann, M., Anbari, F. T., & Bredillet, C. N. (2010). *Perspectives on projects.* London and New York: Routledge.

Zwikael, O., & Globerson, S. (2006). Benchmarking of project planning and success in selected industries. *Benchmarking: An International Journal, 13*(6), 688–700.

17

Increasing Odds in Uncertain Times: Creating Value through Collaborative Learning

Robert E. Bierwolf[1] and Pieter H.A.M. Frijns[2]
[1]Foundation Center of Technology and Innovation Management (CeTIM), Gateway Office, National Organization for Development, Digitization and Innovation (ODI), Ministry of the Interior and Kingdom Relations, Netherlands
[2]Amsterdam Business School, Faculty of Economics, Vrije Universiteit Amsterdam, Amsterdam, Netherlands

INTRODUCTION

Context

Whereas in the past century projects and project management was the exception to the rule (temporary versus standing or permanent organizations), in the last few decades, we have seen the upcoming of what is referred to as "projectification of society." With a clear increase in the last ten years, empirically evidenced, it is not a buzzword but has become part of our reality. In our view, projects are to be seen as vehicles for change as elaborated by a recent PMI in-depth report, emphasizing there is a larger audience and picture to address than shareholders. Placed in the context of what is referred to as our VUCA-world combined with an increase of Digital Transformations implies a further expected growth of that projectification.

Meanwhile, we are facing major societal challenges, with some of these being present daily on the news and also recognized as an emerging field in management (Van Velzen, 2022). Within the public sector, government, many major projects are operating at all of the various levels of government: municipal, provincial, national, and some crossing our borders. As publicly known, not all of our projects run smoothly. A phenomenon that has been referred to for decades as project success and failure and associated Critical Success Factors (Pinto & Slevin, 1987). Given that the number of projects continues to grow, becoming even a human condition as it permeates almost all we do, that also implies a growing need and relevance, for the project success rate to increase (Jensen et al., 2016). And from our perspective, this relates to ever more complex and interconnected societal challenges being addressed through projects which also requires understanding the limitations of what is referred to as conventional project management (Dalcher, 2019). Please note that when we use the singular word/construct project or project management, we implicitly refer to the broader concept of Project-, Program- and Portfolio Management (P3M) (Murray & Sowden, 2015).

Perspective

Our embracement of research in practice does not entail the magical "silver bullet" (a simple solution to a complicated problem), yet does help to increase the chance of success on our projects. To that end, we make use of the Gateway Review™ as a means to support our projects, using what we have coined as the Dutch approach thereto (Frijns, Gommans, et al., 2017). The essential but subtle characteristic is that our use of the method is operated from a learning and helping perspective rather than a controlling and compliance stance. The motivation is major projects should also be perceived as social activities, with the centrality of people as leading thought, that came forth from rethinking and reframing the research agenda (Dalcher, 2019; Winter et al., 2006) and where collaborative learning is a major value in particular in said VUCA context. We seek to re-confirm the value of the "social learning processes of observation, dialogue, storytelling and conversations between people" (Sense & Badham, 2008). And with our history of data available, the Gateway Review Reports as a deliverable of each Gateway Review™ finished, looking back periodically makes sense too, and going forward we gather new data from our contemporary Gateway Review™, thus creating an ongoing (retrospective and prospective) longitudinal case study situation, allowing us to take advantage of the rich empirical data, from which some findings will be presented in this chapter. With these changing research agendas in mind, our key objective of our research and practice is to understand how can we learn collaboratively, and if we can also foster and stimulate creating bridges, for example, bi-directional between research and practice, thus creating a form of learning- or control-loop inducing a "Research through Project Management" approach inspired by the Research through Design paradigm (Bierwolf et al., 2017; Green, 2009).

UNDERSTANDING THE BACKGROUND TO OUR STORY

Related Literature

Considering the need to periodically rethink, in our case rethinking project management and inspired by the research agenda presented in (Winter et al., 2006; Winter & Smith, 2006) as cited in (Bierwolf, 2017b), we underwrite the direction of theory in practice, and supporting the development from practitioners towards reflective practitioners, and understanding that projects are in essence social processes. We align with (Dalcher, 2019) understanding the need to expand from "the pervasive preoccupation with recipes for success, established 'best practices', and standardized Key Performance Indicators (KPIs) and Critical Success Factors (CSFs)." Previously, as also encountered in our practice, too much emphasis on the controlling perspective often leads to risk-avoiding behavior or an anxiety culture, which can hinder work on necessary change needed. We can safely state, despite all project management standards and related research to-date, the (perception of) problems with projects and issues persists.

So what links the phenomenon of projects, their success and failures, with organizational learning and learning organizations, the VUCA context all together? It all manifests itself in our and the typical increasing number of projects run in government or public sector (Hartholt, 2023). In particular, with all the current societal challenges being dealt with, new policy will emerge and the various triggers that will require projects as vehicles for change, e.g., transforming operations, services, performance and can indeed be viewed as a way of organizing and implementing various types of policies (Larsen et al., 2021). For example, such triggers as the inevitable energy transition, impact of climate change and emissions, rippling through into issues relating to the building crises and housing market and city development, generations and demographic changes and labor market, technology changes, ongoing digital transformations are all relevant, requiring to be addressed by policies, invoking many major (innovation) projects (Lidman, 2023).

This also challenges the use of conventional project management methods and tools and invokes the need for augmented approaches and creativity to deal with such (Klein et al., 2015; D. Snowden & Rancati, 2021). Conventional approaches and so-called best practices work effectively in certain conditions as explained in the Cynefin framework (D. J. Snowden, 2000). Such understanding triggers to move towards leveraging the collective intelligence (Conn & Mclean, 2020). Stating that project failures are not uncommon in the practice in the Netherlands when it concerns high-stake programs and major or mega projects at all public levels seems evident. That triggered the use of the Gateway Review™ in the Netherlands and where we build on the collective intelligence of the Gateway Review Community as will be addressed in the next section. To avoid misunderstanding, the Gateway™ Review is neither the subject nor the research method of our message, but foremost our data source.

Gateway Review™ and Dutch Approach

Kicking-off in the Netherlands

In this section, we explain the expanded use related to experiential learning in the Netherlands of the Gateway Review Method (*GRM*) originating from the UK (Office of Government Commerce (OGC), 2007). The method was adopted by the Dutch government in 2009 officially after some initial piloting in the years before triggered by issues in our major projects portfolio, in particular those with a major IT component, but certainly not limited thereto. The role of the Senior Responsible Owner (SRO) as an essential factor was recognized in these reports. Initially the method was adopted for the use at national government level (in Dutch: Rijksoverheid) but was soon expanded with major projects at the municipal level in 2010 and then also served the provincial level in 2011.

Subtle Difference in Mindset

The objective of the Gateway Review™, also in the Netherlands, is to facilitate conscious management decision-making at various gates on whether and how to progress a particular project to ensure the intended benefits or values will be created. These major projects run in fact at all public levels, e.g., national, provincial, and municipal levels, and some also crossing national borders, requiring a longer rather than a short time horizon. More importantly, the Dutch approach uses a helping hand and a learning attitude rather than a controlling and compliance perspective, more of an intrinsically motivated course than an extrinsically motivated one, striving for a positive experience rather than possible negative consequences, using the Gateway Review™ as an opportunity to reflect. This subtle difference has therefore a major impact on the mindset, attitude, behavior, of both the reviewers as well as the reviewees and thus on the outcome of any Gateway Review™ conducted.

Note that the Gateway Review Method (GRM) is just one of many existing tools, instruments, for managing and controlling projects and programs (Bierwolf & Frijns, 2019). The method already has a substantial track record of use and has been in use in the UK since around 2001, as depicted in the "timeline of developments to improve government project delivery" (National Audit Office (NAO), 2016). Whereas Gateway Review Method was adopted over a decade by the Bureau Gateway in the Netherlands and currently, it still operates under the Ministry of the Interior and Kingdom Relations it covers all these type of projects for any of the ministries in the Netherlands, provinces and municipalities. The Bureau Gateway is the only accredited entity in the Netherlands for the Gateway Review™.

Innovation of the Dutch Approach

Over time, within scope of our Dutch approach, some more Gateways were developed as addition to the standard set of Gateways 0–5 as originating from OGC (Bureau Gateway, 2011). These are the so-called "Starting Gate" and "Health Check" with the first one is for moving from a new policy towards defining the relevant program as it obviously depends heavily on the quality of such new policy definition and description.

The so-called "Health Check" which in essence is an organizational scan, and can be performed at any time an organization deems it useful or desirable. As well as an associated "Chain Scan" (in Dutch: ketenscan) to indeed asses the chain of organizations involved, necessary to realize the project (Frijns, Gommans, et al., 2017).

During the COVID-19 pandemic period new different services emerged organically and through piloting (a learn-by-doing manner), with their so-called working titles being "Gateway Review Reflection Day" and the "Gateway Review Dialogue Session" (in Dutch: dialoog tafel).

As to the new service of "Gateway Review Reflection Day", which explicitly does not equal to a standard Gateway Review™, it is maximally two days of interviews with just a handful of key-participants by one or two Gateway Reviewers. The purpose is not to write a similar report with recommendations as comes forth from a normal Gateway Review™. The purpose of a Gateway Review Reflection Day is to give the organization a short and independent reflection on the principles used during a phase transition. This is done on the basis of a limited number of documents and interviews. If desired, the outcomes can form the basis for a true Gateway Reviews further along the process.

The "Dialogue Session" specifically intends to foster the dialogue between all of the stakeholders from the particular program, using the findings from the respective Gateway Review™ and recommendations in the specific Gateway Review Report. and is owned by the Senior Responsible Owner. It augments the classic Senior Responsible Owner evaluation sessions defined in the original OGC Gateway Review™.

Peers and Psychological Safety

In essence, the Gateway Review Method is a peer-review process, but this few word sentence does no justice to crucial elements that provide the foundation that is implied by using the word "peer".

The word "peers" has many implications that need to be well understood and that are conditional for the manner in which the Dutch approach to the Gateway Reviews operates. That is, the Gateway Review Community (GRC) members engage as peers (Noordhoek, 2019) in the temporary Gateway Review Teams (*GRTs*) based on two related but different concepts of "Trust" and "Psychological Safety". Trust is a word that also appears in a Dutch saying implying trust is hard to gain but easy to lose (in Dutch: "vertrouwen komt te voet en gaat te paard"). However, whereas trust is one word, it is not always automatically applicable nor present, and while each person thinks they know what it means or is, actually it deals with broad spectrum of factors (Ruijter et al., 2021; Voortman, 2012).

In our case we refer to peers in particular as in peer-level acceptance and consider it part of and aligning with the social theory of learning, on the other hand we argue that is not just about individual learning but about collaborative learning as part of problem solving learning (Ahern et al., 2014; Wenger, 2007). The underlying phenomenon of (team) psychological safety is one seemingly de facto embraced in the medical sector, but in principle is essential for any sector and organization striving to improve performance using learning (Edmondson, 1999, 2019).

Our Gateway Review Community has been formed over the years and currently exists of some 300+ (past) board members, top level civil servants, program managers, and other accredited Gateway Reviewers. From this community the Gateway Review Teams are assembled for performing the reviews. And from the same community periodically so-called Focus Groups are assembled to debate/discuss particular topics as they emerge, for example based on data analysis.

Do note the implication of all of the foregoing. We are not focusing on the project manager as opposed to the typical discourse does as soon as the wordings project and management emerge, rather we focus on these top level civil servants, who take on the role of the Senior Responsible Owners (SROs). To that end we embrace another typical and almost always reported critical success factor, the engagement of senior management.

Reports as Data

The primary data for our research are the default outputs from the Gateway Review processes. That is, the text data as written in each of the individual Gateway Review Report (GRR), as depicted in (Bierwolf & Frijns, 2019). This accumulates to-date to over 4,000 individual recommendations made in our Dutch case stemming from 500+ reviews involving 10,000+ individual interviews. These reports and their recommendations are structured and stored in a database and periodically analyzed by Bureau Gateway staff. In our case there is no separate typical end-of-project activity required to gather so-called lessons learned as often suggested in literature, and which can be ineffective in collecting and using lessons learned (Buttler, 2016), as our data flows from the standard operating process of conducting our Gateway Reviews.

Embracing Research Methods

A crucial element in the Dutch approach of conducting Gateway Reviews™ is the add-on of using multiple scientific research methods in combination with our aggregated data-set. This was in particular triggered at the first lustrum celebration, on instigation of our Gateway Review Board, wondering if there are any common issues that could emerge from all the reviews (in Dutch: rode draden). This further relates to how we go about it, in all of the associated interviews, meetings, sessions and what comes forth from these. We will keep this section brief. We use *Qualitative Research,* understanding that projects are in essence social processes (Timmermans & Tavory, 2012). We make use of a so called *Community of Practice,* our Gateway Review Community (Kazi et al., 2007). Our community with over 300 Gateway reviewers from all public levels, form a diverse group of people, a group of peers throughout the public service. In the Gateway Review Method, the use of *Triangulation* is embedded, in assessing the situation in the specific project being reviewed, in order to formulate the review team's consensus recommendations for the respective Senior Responsible Owner of that project. We embrace a *Grounded Theory* methodology inspired approach (Kenealy, 2008) that certainly fits management research in combination with the use of *Focus Groups.* Shortly explained, our data, is in principle rich text data. Coding such data, leads to potential patterns to emerge or to be deducted, and with meaning to be added to such patterns Focus Group sessions as previously reported in (Bierwolf & Frijns, 2021), citing (Chaves et al., 2016), which also fits the continuing qualitative and action research approach. The Focus Groups thus interpret these patterns and add a consensus meaning to those, possibly and ultimately leading to a new understanding or insight, followed by new actions or interventions, and adapted action repertoire, rolled out towards the larger group Gateway Review Community, our CoP, and in future Gateway™ Reviews to be conducted.

TO WHAT RESULTS DOES IT LEAD?

In this section, we will provide some more findings and results, some of which have been presented and published before at conferences or in journals, and some that relate to latest emerging insights that may require more/future data and analysis prior to qualifying those as evidenced.

From Patterns to Action Repertoire

Whereas the editors and the authors of (Belack et al., 2019) state they believe that for too long project and program management standards have focused on the "hard skills," these authors next emphasize the growth and importance of "soft skills" which ultimately expands into a complete separate chapter on the subject in their publication. We concur to their position, having provided further independent empirical evidence emerging by placing recommendations that related to elements as Content and Process on the hard-factor side and Culture and Relations on the soft-factor side (Frijns, Van Leeuwen, et al., 2017c).

In the latter article we introduced the Lucky Clover Model (LCM, see Figure 17.1), enabling that such elements as Culture and Relations become consciously part of the mindset all involved in the Gateway Reviews™, triggered by the previously almost sole focus on Content and Process. So while in normal conversations Culture and Relations do emerge, it virtually did not in the set of recommendations.

We applied the Lucky Clover Model as a tool for intervention in Gateway Reviews™, to induce/steer towards a more balanced approach to the projects reviewed and future projects. Note that the purpose to

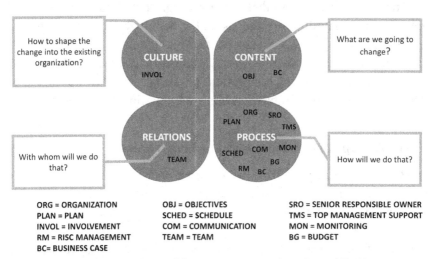

ORG = ORGANIZATION	OBJ = OBJECTIVES	SRO = SENIOR RESPONSIBLE OWNER
PLAN = PLAN	SCHED = SCHEDULE	TMS = TOP MANAGEMENT SUPPORT
INVOL = INVOLVEMENT	COM = COMMUNICATION	MON = MONITORING
RM = RISC MANAGEMENT	TEAM = TEAM	BG = BUDGET
BC= BUSINESS CASE		

Source: P. Frijns, F. Van Leeuwen, and R. Bierwolf, "Project management -A more balanced approach," in *2017 IEEE Technology and Engineering Management Society Conference, TEMSCON 2017*, 2017, pp. p254-258.

FIGURE 17.1
Lucky Clover Model with plotted CSFs.

having an artifact is that it helps guide the discussion and the dialogue. Rather than declaring the model as the goal, it is a means to an end.

The most recent notion is that the Lucky Clover Model's use has been baselined, and it no longer requires active emphasis, and it is a de facto given and starting point. That is, the Lucky Clover Model is considered normal, also with the audience of Senior Responsible Owners requesting reviews, and the member of the Gateway Review Community acting in the respective Gateway Review Teams conducting those reviews, which has become visible in the recommendations where the soft factors are now far more frequently addressed. That shift of focus lead to recommendations that call upon behavioral changes, rather than adherence to process and procedures, facilitating a change in action repertoire.

With the Lucky Clover Model adopted, the previously reported missing balance is indeed established, as soft factors such as "Relationship" and "Culture" have been frequently addressed in the growing set of recommendations of the Gateway Review Reports. The change due to earlier intervention has also changed the Senior Responsible Owner's question formulation at the beginning of a Gateway Review™ to understand his true need. As a reminder, such reviews are mini-projects with a typical twelve-week life cycle from initial contact/question to conducting the review and delivering the report. The questions and reflections that already occur in the early stages, from intake to the planning meeting, assist the Senior Responsible Owner in getting a more precise image of the real question. Similar to the phenomenon, also known as the spiraling process, typically in action research and the research question phrasing, whereby an increased understanding of the problem changes the question (Agee, 2009).

Focus Groups and Project Management Topics

In Bierwolf and Frijns (2020), we reported on Focus Groups (FGs) held on so-called recurring topics (signals) coming forth from the data analysis of the Gateway Review Reports (GRRs) that we will summarize in this section.

Some of those recurring "topics" were picked up as colleagues from the Gateway Review Community stepped forward, based on their personal interest.

Each of these groups addressing a single topic would apply a series of questions to guide the discussion, such as: what or how do I relate to the topic; what is my understanding or interpretation of the topic; what do I see in daily practice related to the topic; what would I like to see. Bureau Gateway staff provided each Focus Group with the relevant data set and patterns from the analysis on these topics, as emerged from the recommendations, as well as some other literature on the topic such as books/articles as reference and to frame the topic. Finally, there is the practical objective, for each group to determine how the various perspectives and insights need to be fed back and applied in the practice of the Gateway Review™.

The aforementioned imbalance in the previous section were categorized using "theme" oriented topics, with five themes to be addressed: [a] Governance, [b] Risk Management, [c] Relationships, [d] Culture, and [e] Leadership.

The findings on [a] Governance are reported in terms of setting the intensity or level of governance against the intensity or level of dynamics involved. This lead to two "types" of governance identified and classified and the wordings chosen to better express their differences as: "Structure Governance" versus "Relation Governance". Each of these were characterized by a different set of words or connotations (Frijns, Van Leeuwen, et al., 2017b).

Related to [b] Risk Management, many of the recommendations of conducted Gateway Reviews™ require SROs to pay attention not just to the adequate management of risks, but require them to actively respond to opportunities. From this Focus Group topic session an additional understanding emerged, relating to a change needed from traditional risk management (on predictable and manageable risks, and a suboptimal use of tools such as risk analysis, risk log and risk reporting) toward dealing with a broad perspective of risks and opportunities, requiring a cultural shift, augmenting the classic risk log towards a risk dialogue (Frijns, Van Leeuwen, et al., 2017d).

On the issue of [c] Relations that emerged, there is the obvious dependency on others and the objective of any project can only be achieved through cooperation (Frijns & Bierwolf, 2018). The Focus Group sessions on this subject provided insight that stakeholders and relationship are not the same. Moreover, conventional stakeholder management does not suffice. Understanding that the language used differs, helps to also adapt the dialogue on a going-forward basis in the Gateway Review™. That is, the current language from and about stakeholders is much about the formal and rational aspects of the collaboration. Whereas the language of relations is more about the informal aspects including emotion and intuition, and subtle intonation or attitude in asking questions by the reviewer during the conduct of a Gateway Review™ can thus have a major impact on the response of the reviewee. Consciously being aware of the social aspects, human aspects and bring them into play has been undervalued and needs to be emphasized on a going-forward basis.

The aforementioned imbalance demonstrated almost absence of attention to [d] Culture, whereas many studies have clarified that culture is Critical Success Factor for the realization of programs and projects (Gemünden et al., 2018). Hence, many a project fails because there is little to no attention for the culture or its alignment with the project. The recommendation of the focus group session stressed the need to consciously discuss any culture issues, much more than was happening.

The outcome of this Focus Group on [e] Leadership is, to each time consciously discuss and figure how the interviewees think and talk about leadership in project under review, which also holds true for the Senior Responsible Owner (SRO). It is about how leadership is shaped and implemented, if it is appropriate for the charter of the project.

All the above further led to intermediate reports or working papers made, which were used for follow-up sessions and discussions in our community. It also impacts and will lead to improvement of the Dutch approach of the Gateway Review™. For example, it implies the course and training to become a certified Gateway Reviewer requires adaption, as well as the selection of new Gateway Reviewers based on mindset. The need for these changes were voiced by the participants of the focus groups themselves, taking ownership.

Latest Development – Wheel of Connection

As we stated above, the use of the Lucky Clover had become 'normal'. Since the introduction of the Lucky Clover Model and addressing the Gateway Review Reports in the related evaluation sessions, the SROs and their teams also expressed what they perceived as important, from which a new set of factors emerged.

Understanding and considering multiple factors that interdependently play a part, that new "artifact" was recently construed, currently named the "The Wheel of Connection," as a successor to our "Lucky Clover." The current working name is Wheel of Connection (WoC), typically came from our Dutch Gateway perspective, underpinning that one needs the collaboration of multiple actors and elements or factors to effectuate the many societal change programs that government faces. The use of a form, shape, of a wheel is the analogy and association to another typical Dutch artifact, the bicycle, and is used to explain and underpin, through metaphor, that none of the elements/spokes of the wheel can operate optimally without the others. The current elements or so-called spokes of the Wheel of Connection Model relate to [a] the societal charter, joint task or challenge/question (in Dutch: de maatschappelijke opgave), [b] connecting to shared and collaborative ownership thereof, an Senior Responsible Owners Association [c] the choice on a specific collaborative configuration/design for the challenge at hand, [d] the management and governance arrangement, [e] selection of the methods and instruments to be used, and [f] how to facilitate the overall approach. The Wheel of Connection Model, of course, has a rim too, representing the program context, and a tire for representing the broader environment at large.

For each of the spokes of the Wheel of Connection Model, a series of related questions emerged. And these also continue to be subject to change as the Wheel of Connection Model is used more and more in practice. First experiences show that these seemingly trivial questions are not always trivial to the respective SROs and their teams.

With the Lucky Clover Model being adopted since its introduction towards currently a "normal state," the next step is emerging. Coming from a focus on professionalizing Project-, Program- and Portfolio Management (P3M) in the governmental setting, through the focus on steering towards "balance," we now move towards facilitating and steering towards "connecting" people in all the engaged organizations and societal groups. Just as Lucky Clover Model is an artifact to support dialogue, and helped craft a new artifact, the Wheel of Connection Model that now emerges. The proliferation, adoption, and impact of the Wheel of Connection Model are expected to materialize similarly as the Lucky Clover Model did over the past years and will be part of the continuing Action Learning, Action Research approach and continuing the use of the approach of Experiential Learning (Learn-by-Doing) in practice, and following the path from reflective practitioners to reflective professionals (Bierwolf, 2017a). The data volume, number of reports, is yet too small to draw firm conclusions other than some early signals that are emerging.

From the limited experience with the Wheel of Connection Model and Gateway Review Reports to date, early signals emphasize fostering the coherent collaboration between these stakeholders in the spirit of the Dutch approach of Gateway Review™, fostering the connection in the spirit of the Wheel of Connection Model. That is, addressing the findings reported in the Gateway Review Report at hand, creating a common perspective and understanding among the participants in the Dialogue Session. By eliciting all their perspectives openly, allowing all to share their narrative, collaboratively construing a common and shared perspective and narrative. Again, such dialogue only works if there is mutual trust, and if participants dare to share their viewpoints and are open to truly listening and understanding the other perspectives, a matter of attitude and mindset. Addressing the Wheel of Connection Model and Dialogue Session have also recently been done through some Review Team Leader's workgroup sessions, to share their experiences to-date with the use of Wheel of Connection Model and Dialogue Sessions. One of the observations or lessons learned relates societal challenge, that can spawn into multiple related programs and projects. It seems to be that perceptions are not about a collaborative social challenge, rather a social challenge requiring collaboration, meaning that each SRO in the SROs Association does not feel or take responsibility for the whole challenge but each does so for their part only.

Once more data is obtained, the same supportive research processes and path will be followed to extract learnings from these new expanded data set, hence the research does not stop.

PERSPECTIVES & CONCLUSION

Projectification and Need to Adapt

The projectification continues and is also relevant in the public sector, but perceiving projects as "the" controllable way of avoiding all the classic problems of bureaucracy with which most "normal" organisations are confronted and struggling with, has proven to be a fallacy (Packendorff & Lindgren, 2014). The mechanism of projects is neither a "silver bullet" nor a panacea.

A key objective that should have become clear from all foregoing sections, is the necessity to accept that in today's projects not all is fully predictable at the outset such project, requiring us to learn as we move along the project life cycle, and to adapt our ways as soon as possible, all with the intent of course, to succeed or more modestly phrased, to increase our chance to success. This mandates us to learn, to learn faster, to learn collaboratively, through periodic reflection. Our use of the Gateway Review™ offers such an opportunity for reflection, inspired and supported by the action research, action learning, experiential learning approach citing (Argyris, 2004; Schon, 1983). Our approach does not deny the value of either Key Performance Indicators (KPIs) and Critical Success Factors (CSFs). But we can step away from the "pervasive pre-occupation" and we can consciously reframed our focus, understanding the necessity to continuously gain new insights based on our continuing data coming forth from our Gateway Reviews™. To ultimately adapt, be it our action repertoire in projects, in the reviews or by adaption our Dutch approach. At same time, we again acknowledge our Dutch approach is certainly not the "silver bullet", nor a panacea. However, we do embrace the position that the chance of success can be increased, if we learn faster, if we adapt faster in the dynamic environment we operate in, and if we thus adapt our behavior and action repertoire in a similar pace.

In the context of Project Based Organizations (PBOs,) the concept of organizational learning is not new, but sometimes not so easy to implement or maintain in a continuously changing environment in particular as it relates to behavior (Argyris, 2004). In the Dutch case, each participating Gateway Review Team member can take new learnings back to their current projects or programs, creating a learning ripple effect, and acting as boundary spanners through socialization and application of acquired knowledge (Brion et al., 2012; Nonaka, 1994). With the common understanding that our ongoing work contributes to what is referred to in the Netherlands as the "learning government/ public sector – in Dutch: lerende overheid", allowing for continuing and ongoing (social) innovation, that is, systemic innovation in both the Dutch approach to the Gateway Review™, as well as in the direction of the reviewed projects and programs.

Through our Dutch approach of the Gateway Review™, we are able to facilitate learning at multiple levels, much in line with the intent of Organizational Development and the development on individual, group or team and organization level, to indeed improve effectiveness (Singh &

Ramdeo, 2020). Be it in our case, as individuals in the projects under review, i.e., a) the respective Senior Responsible Owner and b) interviewed key stake holders; c) each of the participating Gateway Review members of the Gateway Review Team; d) as well as the respective staff of the Bureau Gateway; and e) all these, also in their separate groups or team settings or environments and at the broadest level of the Gateway Review Community; and f) boundary spanning over municipal, provincial, and national level. This aligns with earlier notions of "Individual Learning (IL) and Organizational Learning (OL)" but now as an implemented concept and goes beyond learning in a project (Sense & Badham, 2008). We also gain confidence that this is the right path to follow based on other reportings from another contemporary Community of Practice (CoP) albeit in a different sector (Burgess et al., 2022)

From all the foregoing, we can state that the Gateway Review Community equals the concept of a CoP, and continues to function far beyond the idea of skunk works (Ayas & Zeniuk, 2001). This contrasts with the common notion about governmental, bureaucratically controlled hierarchies and power relationships affecting the projects (Duryan & Smyth, 2019).

Framework of a Knowledge Enrichment Model

To place all presented into a picture that sums it all, we established what we refer to as the Knowledge Enrichment Model (see Figure 17.2) reflecting a more complete learning ecosystem (Bierwolf & Frijns, 2019; Frijns, Van Leeuwen, et al., 2017a).

As we explain in the latter, around 2014, the Bureau Gateway has conducted a perception survey under SROs to experience the Gateway Review Methodology. From this research, SRO's advocated to organize a process that promotes the manifestation of explicit and implicit learning experiences from Gateway Reviews™. This allows for a broader dissemination of results and even better utilization of this in practice. In the same period, a number of Gateway Reviewers encountered the need to discuss the lessons learned from the Gateway Review™ to professionalize program and project management within de Dutch Government. These needs and requests were the trigger for linking structural facts and figures from Gateway Reviews™ to Gateway Reviewers' knowledge and experience. This combination of facts and experience creates new

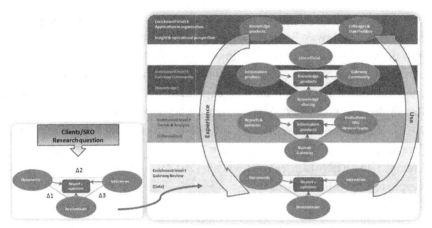

Source: P. Frijns, R. Bierwolf, and F. Van Leeuwen, "Facilitating Learning Organizations An Enrichment Model for Knowledge Management," in 2017 IEEE European Technology and Engineering Management Summit (E-TEMS), 2017, pp. 23–31.

FIGURE 17.2
The Knowledge Enrichment Model.

insights, new knowledge and new perspectives. In order to properly shape this process of knowledge enrichment within the Gateway Community Netherlands, the *Knowledge Enrichment Model* has been developed. This model clarifies how the relationships between the separate Gateway Reviews™ and the content-based lessons are at multiple levels.

The model depicts these levels, visualizes the process of transformation followed from data to information, to knowledge and useable action repertoire, since the Gateway Reviewers are using such insights into their daily practice, and use these experiences in their next assignment in a review team. These are presented (see Figure 17.2) by the two arrows forming a learning circle. That last step is where the higher level value beyond a single Gateway Review™ is realized. which is about using and applying the acquired knowledge and insights into daily practice. There is more than one way how this can be achieved. First, our knowledge products (small booklets in Dutch) are shared among inquiring government officials interested in these lessons learned. Second, the knowledge products are shared with other public organizations aimed at further optimizing the "learning government," such as our training institutes and congress offices and consultancy departments. And to warrant the proper use and alignment to the Dutch approach, these organizations make collaboration arrangements. Thirdly, the learned lessons are used to continuously develop our Gateway Review™ quality and the professionalism of Gateway Reviewers and their training and education. Hence, our model represents the learning ecosystem within our context, but which is generic enough to be replicable in other environments and contexts. That latter thought is expanded in our suggestion coined Research through Project Management (RtPM) approach (Bierwolf et al., 2017), based on the concepts of Research through Design (RtD) (Frayling, 1993), Open Innovation and the European Network of Living Labs (ENoLL) (Evans et al., 2017), and may become more supported by future research.

Concluding Remarks

In this chapter, we addressed how we embrace academic research approaches to the practice of conducting major project reviews, which enables organizational learning and systemic innovation through its community of practice, and span boundaries. This chapter demonstrated the use of what we refer to as the "Dutch Approach" over more than a decade of conducting Gateway Reviews™ in the Dutch government and the concept of "Research through Project Management." We provided insights by sharing our experiential learning from our ongoing longitudinal case study, and hands-on pointers. Ultimately we strive to create a dynamic capability for systemic innovation.

Is there value in our approach? A first clear signal about the value creation is the positive valuation in our case by the SROs requesting Gateway Reviews™, which was captured in the 2019 conducted series of interviews/surveys with a subset of 16 of the SROs. And is further supported by many more SROs and their continuing use of our services. The overall conclusion is that all SROs are convinced of the added value of the Dutch Approach to Gateway Review™ and that the continuous development, innovation, contributes to the learning capacity of the Dutch government (Edelbroek & Steensma, 2019).

And as a last note to the practitioners that have become curious after reading our chapter, note that the section "References" is not only provided for justification, but also and specifically for those management practitioners to whom a particular paragraph struck as interesting, to stimulate a snowballing effect and open mindset, to seek further reading supporting the pursuit (Sense & Badham, 2008) of lifelong learning (Bierwolf, 2017a).

REFERENCES

Agee, J. (2009). Developing qualitative research questions: A reflective process. *International Journal of Qualitative Studies in Education*, *22*(4), 431–447.

Ahern, T., Leavy, B., & Byrne, P. J. (2014). Knowledge formation and learning in the management of projects: A problem solving perspective. *IJPM*, *32*(8), 1423–1431.

Argyris, C. (2004). Double-loop Learning and Organizational Change Facilitating Transformational Change. In J. J. Boonstra (Ed.), *Dynamics of Organizational Change and Learning* (pp. 389–401). John Wiley & Sons, Ltd.

Ayas, K., & Zeniuk, N. (2001). Project-based learning: Building communities of reflective practitioners. *Management Learning*, *32*(1), 61–76.

Belack, C., Di Filippo, D., & Di Filippo, I. (2019). *Cognitive Readiness in Project Teams Reducing Project Complexity and Increasing Success in Project Management* (C. Belack, D. Di Filippo, & I. Di Filippo (Eds.); 1st ed.). Productivity Press, Routeledge, Taylor & Francis Group.

Bierwolf, R. (2017a). Practitioners, reflective practitioners, reflective professionals. *IEEE EMR*, *45*(2), 19–24.

Bierwolf, R. (2017b). Towards project management 2030 – Why is change needed? *IEEE EMR*, *45*(1), 21–26.

Bierwolf, R., & Frijns, P. (2019). Project Management – Towards a More Complete Learning ecoSystem? *2019 Int. Con. on Engineering, Technology and Innovation (ICE/ITMC)*, pp. 1–6.

Bierwolf, R., & Frijns, P. (2020). Digital Transformations and Learning Communities in the Practice of Project Management. In L. Pretorius & M. W. Pretorius (Eds.), *29th Int. Con. of the Int. Association for Management of Technology (IAMOT 2020)* (pp. 495–507).

Bierwolf, R., & Frijns, P. (2021). Governance – Structures and human perspectives. *IEEE EMR*, *49*(2), 172–180.

Bierwolf, R., Romero, D., Pelk, H., & Stettina, C. J. (2017). On the Future of Project Management Innovation: A Call for Discussion Towards Project Management 2030. *2017 Int. Con. on Engineering, Technology and Innovation (ICE/ITMC)*, pp. 711–720.

Brion, S., Chauvet, V., Chollet, B., & Mothe, C. (2012). Project leaders as boundary spanners: Relational antecedents and performance outcomes. *IJPM*, *30*(6), 708–722.

Bureau Gateway. (2011). *OGC Gateway Review Producten & Diensten Catalogus* (Nov 2011). Ministerie van Binnenlandse Zaken en Koninkrijksrelaties, De Werkmaatschappij, Bureau Gateway.

Burgess, N., Currie, G., Crump, B., & Dawson, A. (2022). Leading change across a healthcare system: How to build improvement capability and foster a culture of continuous improvement. In *Report of the Evaluation of the NHS-VMI partnership* (Issue Spring).

Buttler, T. (2016). *Collecting lessons learned: How project-based organizations in the oil and gas industry learn from their projects.*

Chaves, M. S., Cristina, C., Araújo, S. De, & Rosa, D. V. (2016). A new approach to managing Lessons Learned in PMBoK process groups: The Ballistic 2. 0 Model. *International Journal of Information Systems and Project Management*, *4*(1), 27–45.

Conn, C., & Mclean, R. (2020). Six problem-solving mindsets for very uncertain times. *McKinsey Quarterly, September* 1–7.

Dalcher, D. (2019). *Leading the Project Revolution – Reframing the Human Dynamics of Successful Projects* (D. Dalcher (Ed.); 1st ed.). Routledge.

Duryan, M., & Smyth, H. (2019). Cultivating sustainable communities of practice within hierarchical bureaucracies: The crucial role of an executive sponsorship. *IJMPB, 12*(2), 400–422.

Edelbroek, L., & Steensma, L. (2019). *Wat leren opdrachtgevers?* (Issue september).

Edmondson, A. C. (1999). Psychological safety and learning behavior in work teams. *Administrative Science Quarterly, 44*(2), 350.

Edmondson, A. C. (2019). *The Fearless Organization: Creating Psychological Safety in the Workplace for Learning, Innovation, and Growth.* John Wiley & Sons Inc.

Evans, P., Schuurman, D., Ståhlbröst, A., & Vervoort, K. (2017). *Living Lab Methodology Handbook* (K. Malberg & I. Vaittinen (Eds.)). U4IoT Consortium.

Frayling, C. (1993). Research in Art. *Royal College of Art Research Papers.*

Frijns, P., & Bierwolf, R. (2018). Relationship Management – The Heart of Project- and Program Management. *2018 Int. Con. on Engineering, Technology and Innovation (ICE/ITMC),* pp. 321–325.

Frijns, P., Gommans, J., & Leeuwen, F. van. (2017). *Gateway Review in Nederland: The Dutch Approach* (1st print). Ministerie van Binnenlandse Zaken en Koninkrijksrelaties, Uitvoeringsorganisatie Bedrijfsvoering Rijk, Bureau Gateway.

Frijns, P., Van Leeuwen, F., & Bierwolf, R. (2017a). Facilitating Learning Organizations – An Enrichment Model for Knowledge Management. *2017 IEEE Technology and Engineering Management Summit (E-TEMS),* pp. 23–31.

Frijns, P., Van Leeuwen, F., & Bierwolf, R. (2017b). Governance – What Governance. *2017 Int. Con. on Engineering, Technology and Innovation (ICE/ITMC),* pp. 240–244.

Frijns, P., Van Leeuwen, F., & Bierwolf, R. (2017c). ProjecT Management – A More Balanced Approach. *2017 IEEE Technology and Engineering Management Society Conference, TEMSCON 2017,* pp. 254–258.

Frijns, P., Van Leeuwen, F., & Bierwolf, R. (2017d). Risk Management – From Risk Log to Risk Dialogue. *2017 Int. Con. on Engineering, Technology and Innovation (ICE/ITMC),* pp. 721–725.

Gemünden, H. G., Lehner, P., & Kock, A. (2018). The project-oriented organization and its contribution to innovation. *IJPM, 36*(1), 147–160.

Green, L. W. (2009). Making research relevant: If it is an evidence-based practice, where's the practice-based evidence? *Family Practice, 25*(SUPPL. 1), 20–24.

Hartholt, S. (2023). *Overzicht: zo staan de tien duurste IT-projecten van de overheid ervoor.* AG Connect.

Jensen, A. F., Thuesen, C., & Geraldi, J. (2016). The projectification of everything – project as a human condition. *PMJ, 47*(3), 21–34.

Kazi, A. S. (Sami), Wohlfart, L., & Wolf, P. (Eds.). (2007). *Hands-On Knowledge Co-Creation and Sharing: Practical Methods and Techniques.* KnowledgeBoard.

Kenealy, G. J. J. (2008). Management Research and Grounded Theory: A review of grounded theorybuilding approach in organisational and management research. *The Grounded Theory Review, 7*(2), 95–117.

Klein, L., Biesenthal, C., & Dehlin, E. (2015). Improvisation in project management: A praxeology. *IJPM, 33*(2), 267–277.

Larsen, A. S. A., Volden, G. H., & Andersen, B. (2021). Project governance in state-owned enterprises: The case of major public projects' governance arrangements and quality assurance schemes. *Administrative Sciences, 11, 66*(3), 1–27.

Lidman, L. (2023). The gap between the rhetorical why and the practical what and how of public sector innovation. *International Journal of Public Administration, EarlyOnline*, 1–11.

Murray, A., & Sowden, R. (2015). *Introduction to Portfolio, Programme and Project Management Maturity Model (P3M3)* (p. 67). AXELOS Ltd.

National Audit Office (NAO). (2016). *Delivering major projects in government: A briefing for the Committee of Public Accounts.*

Nonaka, I. (1994). A dynamic theory of organizational knowledge creation. *Organization Science, 5*(1), 14–37.

Noordhoek, P. (2019). *Trusting Associations: A Surgent Approach to Quality Improvement in Associations.* "Meer d>n Nu."

Office of Government Commerce (OGC). (2007). *OGC Gateway^{TM} Process Review 0: Strategic Assessment* (v2 ed.). www.opsi.gov.uk.

Packendorff, J., & Lindgren, M. (2014). Projectification and its consequences: Narrow and broad conceptualisations. *South African Journal of Economic and Management Sciences, 17*(1), 7–21.

Pinto, J. K., & Slevin, D. P. (1987). Critical factors in successful project implementation. *IEEE Transactions on Engineering Management, EM-34*(1), 22–27.

Ruijter, H., van Marrewijk, A., Veenswijk, M., & Merkus, S. (2021). "Filling the mattress": Trust development in the governance of infrastructure megaprojects. *IJPM, 39*(4), 351–364.

Schon, D. A. (1983). *The Reflective Practitioner – How Professionals Think in Action.* Basic Books Inc.

Sense, A. J., & Badham, R. J. (2008). Cultivating situated learning within project management practice: A case study exploration of the dynamics of project-based learning. *IJMPB, 1*(3), 432–438.

Singh, R., & Ramdeo, S. (2020). *Leading Organizational Development and Change - Principles and Contextual Perspectives* (1st ed.). Palgrave Macmillan.

Snowden, D. J. (2000). The Social Ecology of Knowledge Management - Cynefin: A Sense of Time and Place. In C. Despres & D. Chauvel (Eds.), *Knowledge Horizons: The Present and the Promise of Knowledge Management* (pp. 237–265). Butterworth-Heinemann, Elsevier Inc.

Snowden, D., & Rancati, A. (2021). *Managing Complexity (and Chaos) in Times of Crisis - A Field Guide for Decision Makers Inspired by the Cynefin Framework.* Publications Office of the European Union.

Timmermans, S., & Tavory, I. (2012). Theory construction in qualitative research: From grounded theory to abductive analysis. *Sociological Theory, 30*(3), 167–186.

Van Velzen, J. (2022). *Gesteund opgavegericht werken* [Univeristeit Utrecht].

Voortman, P. M. (2012). *Vertrouwen Werkt - Over werken aan vertrouwen in organisaties* (1st ed.). Trustworks.

Wenger, E. (2007). *Communities of Practice: Learning, Meaning and Identity* (15th ed.). Cambridge University Press.

Winter, M., & Smith, C. (2006). *Rethinking Project Management – Final Report.*

Winter, M., Smith, C., Morris, P., & Cicmil, S. (2006). Directions for future research in project management: The main findings of a UK government-funded research network. *IJPM, 24*(8), 638–649.

Index

Printed in the United States
by Baker & Taylor Publisher Services